# THE SURVIVAL IMPERATIVE

# THE SURVIVAL IMPERATIVE

Using Space to Protect Earth

## William E. Burrows

A TOM DOHERTY ASSOCIATES BOOK  NEW YORK

A Forge Book
Published by Tom Doherty Associates, LLC
175 Fifth Avenue
New York, NY 10010

www.tor-forge.com

Forge® is a registered trademark of Tom Doherty Associates, LLC.

Library of Congress Cataloging-in-Publication Data

Burrows, William E., 1937–
    The survival imperative : using space to protect Earth / William E. Burrows.
        p. cm.
    "A Tom Doherty Associates book."
    ISBN-13: 978-0-7653-1115-3
    ISBN-10: 0-7653-1115-1
    1. Astronautics, Military.  2. Artificial satellites.  3. Outer space—Civilian use.
I. Title.

PS3602. U769S87  2006
813'.6—dc22

                                                        2005033803

First Hardcover Edition: August 2006
First Trade Paperback Edition: July 2007

0  9  8  7  6  5  4  3  2  1

To Sophie, Tommy, and the other children.
And to the protection they deserve.

# CONTENTS

# PREFACE

Russell L. "Rusty" Schweikart, a veteran of the March 1969 Apollo IX mission, led a session at a four-day planetary defense conference in Garden Grove, California, in late February 2004. Schweikart has a long-standing interest in asteroid-Earth collisions and is the founder and president of the B612 Foundation, which addresses the asteroid threat. Schweikart was in the audience when a speaker at another session noted, correctly, that a city-busting impactor that could kill many millions of people hits this planet on an average of once every several thousand years.

The laconic former astronaut took the point and then answered with typical terseness. "That could be tomorrow."

Indeed it could. This book was written to accomplish two broad objectives. One is to acquaint its readers with the nature of asteroids and comets as they relate to endangering Earth, however remotely. The second is to describe the wide spectrum of attitudes that people hold about a serious or catastrophic collision. One end of that spectrum is occupied by professional and amateur experts, certainly including astronomers, who take a hard look at the probability and likely effects of many kinds of impacts and who try to devise rational means to prevent them, or if that fails, to minimize the destruction. The overwhelming majority of the human inhabitants of Earth neither know about, nor particularly care about, an asteroid or comet crashing into their planet. They are necessarily preoccupied with more mundane matters. The other end of the spectrum is occupied, successively, by those who are fatalistic about a major collision, and finally by those who believe that this world's fiery end is predicted in both testaments of the Bible, and that when Armageddon does happen, true believers will find rapture with Christ in heaven, leaving the others to head in another direction.

Indeed, people's reaction to possible obliteration, and what they would do in the event of a major collision are, to my mind, far more interesting than the consequences of the catastrophes themselves. This is therefore really a story about people and the survival or extinction of the human race, not about space technology and large dirty rocks.

# AUTHOR'S NOTE

The first chapter is an invention. It describes a near-worst-case scenario that is statistically remote but possible. As the chapters that follow show, the sequence of events that occur as a result of the asteroid impact—a nuclear war on the Indian subcontinent, the worldwide communication and transportation meltdown, and the global environmental devastation that result from them—are all plausible in the view of many scientists and others who have thought about the consequences of a collision with a large asteroid or comet. So is the worst-case scenario: complete annihilation. My purpose is not to foretell a worldwide catastrophe, however. It is to prevent it.

# THE SURVIVAL IMPERATIVE

# : 1 :

# HELL ON EARTH

The Rogue came out of the Oort Cloud almost 2 million years ago, and by the time it sped past Earth, its velocity was twenty-six miles a second. Then the Sun's gravity took hold and its speed increased to nearly four hundred miles a second. As it swung around the far side of the Sun, the scorching heat broke it into a six-hundred-mile-long string of huge rocks, ice, and other cosmic debris. When it came out from behind the Sun, the string headed right back to Earth. It got there twenty-nine days later.

The string was finally spotted seventeen hours, nine minutes, and forty-two seconds before it began impacting, so there was no real warning time (as if it would have mattered). It was picked up almost simultaneously by an observatory in Australia and an incredulous and thoroughly terrified seventeen-year-old amateur in Japan.

Spaceguard, which was supposed to spot such intruders and give plenty of warning time, alerted NASA, which alerted the White House, which sent the warning to the U.S. Strategic Command at Offutt Air Force Base in Nebraska. The command's mission was to protect the country from threats in space. But the threats were defined as man-made and consisted mainly of long-range ballistic missiles and weapons launched by enemies that wanted to knock out American spacecraft. It had been accepted during the Cold War that the Soviets would have "taken out" U.S. reconnaissance and ballistic-missile early-warning satellites as a prelude to all-out war because doing so would effectively have made the nation's defense blind. That kind of threat could be dealt with by attacking the enemy in space or on the ground. After all, a weapon conceived by men could be defeated by them. But there was no possible defense against a string of giant rock and ice fragments as long as the width of France that now slowed back down to twenty-six miles a second and appeared with very short notice.

The military was put on Defense Condition 1, the highest level of alert, but of course that was a charade. Meanwhile, the National Security Council hurriedly met and debated whether to tell the people of the United States, and therefore the world, that the string was heading right for them at searing velocity. The council also discussed warning India, Pakistan, China, and Russia. It was decided not to tell the public for fear of starting a panic that would cause chaos and probably political and social disintegration. But the men and women who sat around the long table in the White House decided to quietly warn the Russians and the Chinese.

The men in the Kremlin and in the Forbidden City were of two minds about the news. Some thought it was a diabolical trick designed to destabilize their countries by causing mass hysteria. Others suggested that the dire warning masked an impending nuclear attack ordered by the archconservatives in the White House. So the Russian and Chinese militaries, including their strategic missile forces, were also put on hair-trigger alert. That information was quickly sent to Washington. A deterrent is useless, after all, unless the opposition knows about it.

Moscow University's large telescope was down for repairs, but the one at Zlatoust, high in the Urals, was trained on the string as soon as the alert came from Moscow. The shaken astronomers at Zlatoust spotted it right away and reported what they saw to their leaders. But it wasn't going to matter.

The string streaked past the Moon. Then, at a little before eight in the evening, it came in high over northeastern Australia, causing a clap of thunder that smashed windows. The rocks were now glowing crimson because of atmospheric friction. Then they began to break into fiery chunks, some as small as fifty meters, as they plowed deeper into the atmosphere over Southeast Asia.

One huge icy rock exploded like fireworks six miles over Peshawar in northern Pakistan. The place had become notorious in early May 1960 when it was revealed that Francis Gary Powers, the American U-2 pilot who was shot down over the Soviet Union, had taken off from the air base at Peshawar. The facility was still active when the first rocks in the string came in. There were sixty-seven military aircraft at Peshawar that day whose mission was to stop Indian bombers before they reached Pakistani targets. The exploding rock showered the base with hot fragments and disappeared in a blinding flash. Most of the planes, structures, and vehicles were destroyed, and there were scores of dead and wounded.

The United States knew what was going to happen at Peshawar even before the explosion. Technicians routinely monitored asteroids and comets with optical, X-ray, dosimeter, and other sensors carried on spacecraft and with ground-based telescopes and other equipment in a Nuclear Detonation Detection System that had been developed during the Cold War. The system was created to provide near-real-time information on the enemy's nuclear weapons programs and to verify compliance with the Limited Test Ban and Non-Proliferation treaties. But Pakistan had no such system. Its early-warning radars, which were crude by U.S. and Russian standards, interpreted that first exploding rock as an attempted first nuclear strike on an important military airfield by the despised Indians. Most of the scientists, politicians, and military officers in Pakistan's Nuclear Command and Control Authority conferred immediately on an emergency conference line and advised the president to retaliate before more targets were hit and the nation became too crippled to strike back. All sixteen of Pakistan's long-range Taepodong nuclear ballistic missiles were therefore launched at preassigned targets in India, including New Delhi and Bombay, seven air bases, and what were thought to be missile installations. The Taepodongs were from North Korea and had been traded for Pakistani nuclear-weapons technology, which had in turn come from China.

Still another fragment the size of a small house struck Korolev, the site of the Russian space operations center, on the other side of the beltway eighteen miles northeast of central Moscow. The town was pulverized and the space operations complex, including the mostly gone-to-seed Yuri A. Gagarin Cosmonaut Training Center, effectively disappeared. So the Russian space program was now dysfunctional, its communication links broken and most of its best engineers and administrators dead. Several other towns in the area were severely shaken, and airplanes at two regional airports were knocked to their bellies. Most of Moscow itself was spared. But other large chunks impacted along the steppes, some turning whole towns into smoking craters, others crashing harmlessly into farmland and meadows.

Another of the string's rocks, this one one hundred meters wide, streaked toward St. Petersburg. The shock wave it made as it split the thickening air immediately before impact knocked down every structure and living thing along a seventy-mile-wide swath of countryside as if they had been caught in a titanic earthquake. An instant before it impacted, it turned the evening sky brighter than the Sun. The noise was literally deafening.

The rock struck as the city's symphony orchestra was starting to play the allegro molto of Rachmaninoff 's second symphony. Those in the Grand Hall in the Philharmonia on Mikhailovskaya Street, like most of the city's 4 million souls, never knew what hit them. If they saw and heard the instrument that ended their existence, it was in the barest fraction of a second. They were lucky. Many thousands of others would suffer terrible burns or be severely crippled, some for the rest of their dreadful lives. Three centuries of culture was vaporized in as many seconds, first by the shock wave, and then by the impact itself. The venerable Peter and Paul Fortress, where czars were buried, the Hermitage, the baroque splendor of the Summer and Winter palaces, St. Isaac's Cathedral with its famous golden dome, Aleksandr's Column, the painstakingly restored thousand-room Konstantinovsky Palace and the glittering Amber Room in the Catherine Palace, the Pushkin, Dostoyevsky, and Pavlov museums and libraries, the university and the public library, the Repins and other exquisite paintings, all of the other cultural artifacts and symbols of baroque grandeur, the Neva canals, bustling Nevsky Prospekt, the houses, and the millions of men, women, and children who dwelled in them were instantly pulverized into dust particles that swirled and tumbled skyward in a vast, dark, roiling column of particulate debris. The cradle of Russian culture, the city Pushkin immortalized in verse, the "Venice of the North," orphaned by the Communists who renamed it for their own saint, and then resurrected after their demise, literally no longer existed. That first rock punched a seven-mile-wide hole in the Earth's hide and disintegrated there, along with every vestige of its target. As the rest of the string impacted along a lengthening line that moved westward as the planet rotated, across Northern Europe, the Atlantic, the United States, and Canada, the Earth repeatedly shuddered in convulsions.

Because St. Petersburg was on the Gulf of Finland, the impact also started a 130-foot-high tsunami that swept over the Finnish coast and inundated Helsinki. It kept moving in a colossal wave that struck Stockholm and much of the Swedish coast, battering everything in its path and drowning uncounted thousands.

Tons of earth and debris rose into the upper atmosphere and formed an immense dark cloud of thick dust that soon blanketed Eurasia and started moving eastward, carried by the wind. Torrential rains of caustic acid began to fall from the churning clouds, poisoning the land they soon covered. If sunrise marked morning, there would be no morning.

All five boroughs in New York were severely damaged by a salvo of speeding rocks, and so were its suburbs, as well as other cities in the area from Newark to New Haven. Roughly 3.5 million people were squashed, killed by overpressure, flying debris and glass that struck like shrapnel, or were instantly broiled alive by the heat. Thousands were lifted off streets, blown out of buildings, and were themselves turned into missiles by five-hundred-mile-an-hour winds that smashed them into vast mounds of rubble, either killing them instantly or leaving them all but dead with broken bones, fractured skulls, and torn, twisted, and hemorrhaging insides.

What happened to St. Petersburg happened to a greater or lesser extent all along the string's impact path. Many of the best research universities in the world, untold millions of books and databases, manuscripts, irreplaceable works of art, exquisite architecture of every description, national monuments including the Statue of Liberty, the second World Trade Center, vaccines, medical supplies, and countless victims disappeared in tumultuous storm clouds that soon also darkened the sky.

The North American Aerospace Defense Command's early-warning radar system, the most sophisticated in the world, saw the line of rocks coming like a cracking whip. But the men and women secreted behind a massive steel door deep in Iron Mountain in Colorado Springs had been taught to deal with bomber and ballistic-missile threats and not, as some of them would have put it, with acts of God. So the inhabitants of Iron Mountain and the missileers saw nothing to indicate that a full-scale ballistic missile attack was under way. They therefore stayed hidden behind the great door and in the underground silos and waited.

Not so in India. When the first of the Pakistani warheads struck their targets, turning them into the legendary mushroom clouds, the Indians responded in kind. All forty-two of their nuclear missiles were launched at predetermined targets, including Peshawar. Seeing that their old rival had been severely wounded by the Pakistanis and had then launched a reprisal attack against them, Beijing's rulers decided to take the occasion to destroy India once and for all and launched their own limited, but immensely destructive, attack. By dawn—or what would have been dawn—large areas of India and Pakistan were burning and were in a death shroud of intense radioactive smog. And that lethal smog was already on the move. There were dead in the many millions; no one knew how many because those who ordinarily counted the dead and wounded after disasters were dead, too. India and

Pakistan effectively ceased to exist as political entities. Kashmir, the catalyst of so much sectarian hatred and death on both sides for so long, no longer mattered and never would again.

The vast majority of those on both sides who survived the multiple impacts quickly panicked as chaos took hold. What had once been cohesive communities quickly disintegrated into dazed individuals and small groups that began scavenging and looting. Most civil servants abandoned their offices and blended with the others who wandered through the darkened streets. To their credit, most of the police and the military tried to stick to their jobs, at least in the beginning, and tried to prevent total anarchy. But they couldn't. On the subcontinent, most surviving Indian and Pakistani physicians and nurses treated as many of the injured—hundreds of thousands of them critically—as best they could until medical supplies were gone. That didn't take long. Nor did it take long for the food and water shortages to turn once law-abiding citizens into armed looters. Within twenty-four hours, as the enormity of the catastrophe began to take hold, the social fabric in and around the destroyed regions quickly unraveled. As healthy survivors began to grasp the fact that food and shelter were not only decimated, but poisoned, they became desperate to survive and keep their families alive. And so fighting started, some of it with guns, as competition for food, clothing, and shelter increased. And it wasn't only about necessities. Everything was fair game, including computers, radios, televisions, and any furniture that could be carried. As gunfire broke out in cities and suburbs, the dark clouds blanketed the sky, shutting out most sunlight over much of the planet. The clouds drifted eastward on currents of air in the upper reaches of a poisoned atmosphere. The stench of millions of unburied dead was terrible, and so was the threat of disease they bequeathed to the living.

Except for the radiation, which was spreading, it was the same across a broad swath that extended from western Russia through Poland, northern Germany, Belgium, northern France (narrowly missing Paris), the Channel, the south of England, the North Atlantic, and across the northern United States and southern Canada. Scores of rocks punched harmlessly into the ocean and the Great Lakes and impacted in forests, farmlands, and other sparsely populated areas on both sides of the U.S.-Canadian border. The carnage and destruction were far more severe in Europe, of course, since it was more densely populated. But Buffalo, Toronto, New York, Chicago, Detroit, Philadelphia, and Milwaukee were not so lucky. They took some very bad hits.

The fate of the leaders of the countries under the barrage of rocks was not much better than that of ordinary citizens. Following accepted practice, the notables and their apparatchiks had taken long-standing measures to assure their own survival. Or so they thought. This was invariably justified in the cause of patriotism, since leaders were responsible for holding their countries together, and they obviously could not do that if they were gravely injured or dead. Besides, high office necessarily brought privileges, the most important of which was a head start on survival. So in Russia, the Council of Federation, its president, the entire Duma, and the heads of important ministries, including defense, communication and information, the economy, foreign affairs, and transportation, a few top generals and admirals, and whatever well-placed bureaucrats could be found and warned, were quickly evacuated to several fully stocked underground shelters that had been dug deep along the Moscow beltway early in the Cold War. Europeans, who had been given very short warning time, and who for the most part had no such shelters, hid in their basements.

The president of the United States, members of the National Security Council, and other senior civilian and military advisers, like their foreign counterparts, hurriedly fled Washington in helicopters to a safer place. It was a complex of relatively comfortable shelters that had been carved out of the West Virginia mountains during the Cold War for the same reason the Soviets had prepared to hide underground.

Most of the teachers and other civil servants, as well as doctors and nurses in or near the devastated American cities, kept their heads and behaved valiantly, at least during those first days. Marion Elbert, a bright and self-possessed twenty-six-year-old public-school teacher, tried to get her traumatized fourth graders to the basement of P.S. 139, an ancient, though renovated, brick building on Sixty-third Drive in Rego Park, Queens. The school was half a mile from an impact where Woodhaven Boulevard and the Long Island Expressway had converged, and that was now a smoldering hole. The explosion was close enough to shatter the three large windows in her classroom and bring down part of a wall with watercolor pictures of the animals in the Central Park Zoo that had been made on a field trip. Seeing that other walls were badly cracked, and worrying about the ceiling collapsing, Miss Elbert decided to move her terrified and dazed ten-year-olds into the school yard to wait for whichever parents managed to pick their way through the wrecked neighborhood to find their children. Elbert was as steadfastly protective of the children as if they were her own sons and daughters. She had sense enough to tell them to bring their still-closed pint-size milk containers, but not the oatmeal cookies and apples, which were flecked with tiny fragments of glass and dusted with asbestos. She calmly sat them down on one of the basketball courts and told them to pretend they were in a hurricane movie. Norma Potter, Meg McNally, and the other teachers tried to protect their students, too. So did thousands of other teachers in the region and in other demolished areas.

Some survivors made their way to the houses of worship that were still standing to find whatever comfort they could there. And the few priests, pastors, rabbis, and other spiritual leaders who talked with congregations tried to calm them and ease their shock and anguish. There was the consolation, some of the holy men told their parishioners, that what had happened to them was the Lord's will. It was some kind of a test, they said, reassuringly. Elsewhere in the country, out across the vast, flat, fertile plains of the Midwest and as far south as the Rio Grande, some shepherds told their flocks that what God had inflicted on their countrymen back East was punishment for their sins, just as the Reverend Jerry Falwell had done within hours of the attack on the World Trade Center on September 11, 2001. (Falwell had quickly apologized.) And the sheep believed it. It was a land where more than half of the people believed in creationism, not in evolution, which they dismissed as "junk science." But most of the holy men concentrated on survival, not blame, and searched for constructive ways to survive the calamity.

Most hospitals and other health facilities in and around the devastated areas were effectively destroyed. The physicians, nurses, and other health care professionals were as calm and eager to help as the police and teachers, at least in the first hours. But medical supplies were quickly exhausted. Many of them had been stolen.

The National Command Authority, which now included a president who was hiding in a cave, decided that massive help should be sent to the stricken cities, but that some had to be held in reserve. Sending massive amounts of aid to the devastated areas quickly was not only humane, they agreed, but was the proper role of government. The closest the country had come to experiencing devastation on this scale within a century was when Hurricane Katrina had struck the Gulf Coast at the end of the summer of 2005. There was confusion then about how to help, so federal aid was late, and that contributed to more than a thousand deaths and untold suffering.

But there was a problem. The men and women in the commodious cave also knew that the danger to the United States extended well beyond the cities that had been impacted. They had been briefed on what had happened to St. Petersburg, as well as on the nuclear war that had flared between the Indians and the Pakistanis, and they therefore knew what the dark gray sky meant. The American astronomer Carl Sagan and four colleagues had predicted many years earlier that debris in the upper atmosphere following an all-out nuclear war between the superpowers would be so thick and tenacious, so persistently opaque, it would keep sunlight off the planet and bring on a long, bitter cold, nuclear winter. There had not been an all-out nuclear war, of course, but the net effect of the asteroid impact and the war on the subcontinent was nearly the same. That meant a time of great hardship, perhaps threatening the integrity of the nation itself, was almost undoubtedly at hand. And if that was the case, the reasoning went, government's highest obligation was to fortify as best it could the large part of the nation—certainly including its heartland—that remained unscathed. Help for the stricken cities and other places therefore had to be parceled prudently.

Throughout the piles of smoking or burning debris that had been apartment buildings and offices, schools and libraries, shops and factories and private homes, wandered the walking dead. Those with third-degree burns had pain only on the edges of the wounds because the intensity of the heat had destroyed the nerve endings in the injured areas themselves so the pain could not be transmitted. Their skin looked scalded, charred, or melted, and it would be scarred for whatever remained of their wretched lives.

The inhabitants of what had been St. Petersburg and its surroundings as far as Novgorod, Chudovo, and Nyandoma roamed, injured and dazed, through the vast remains of their own hell. They would not have been consoled knowing that at least there was no radiation. (It was on the way.) The invisible poison and the sky that never cleared were going to blanket almost all the planet.

The destruction of the World Trade Center in New York on September 11, 2001, had brought out the best in human nature. People caught in an unprecedented calamity not only helped each other, but gave their lives to rescue others. Yet as horrible as that attack had been, it was local, not regional. Survivors could therefore be taken to safety and treated nearby, and help could be brought to the scene. But there was a profound difference between the destruction of the Twin Towers and the Rogue's impact on the whole region. People near the towers could walk ten blocks and buy a sandwich or a pizza. That was definitely not the case when the impactors struck.

As happened in the New Orleans area in the aftermath of Katrina, the realization that virtually the entire region was devastated quickly led to chaos as competition for precious resources began. Faced with their own exposure and starvation and the deaths of their children and other loved ones, millions of survivors in the stricken regions went on rampages for food, water, clothing, and shelter. Looting broke out within hours of the impacts. Armed bands and individuals roamed through the debris scavenging for sustenance. And not just sustenance. The looters also carried off television sets, radios, DVD players, iPods, and other appliances, as well as computers, and high-end clothing, from jeans to furs and designer clothes and shoes. Many who were desperate to protect what was left of their families attacked each other with anything that could be used as a weapon, including rocks, bricks, baseball bats, and shovels. At the same time, gunfire broke out almost everywhere, by no means all of it because of the imperative to survive. Snipers behind open windows and on roofs randomly picked off fellow citizens because of some vague urge to protect their neighborhoods, or to settle old grudges, or simply to commit unrestrained murder. Others fired at police and national guardsmen, and the fire was returned. But that didn't last long because increasing numbers of the police and soldiers began to desert with their own weapons and large stocks of munitions.

The crater that had been St. Petersburg was still smoldering—still sending up a thick column of particulate matter like some volcano of nightmarish proportion—two weeks after the Rogue impacted. The remains of the Russians and their city, now reduced to minuscule flotsam that floated upward, soon moved around Earth until the planet was overcast everywhere except in the far northern and southern reaches. Combined with the cesium, tritium, and other radioactive elements the Indians and Pakistanis had contributed to the environment, the remains of the city Peter the Great had built to celebrate the nobility of Russia became a toxic mantle of soot that enveloped Earth.

Now, thanks to the random explosion of a giant star 2 billion years earlier and two tribes of deadly enemies that were fixated on exterminating each other in the name of their gods, it was late autumn, if not winter, under an earthen sky.

The Russians, particularly those in Siberia, were the first to feel the effect of constant cloud cover. Like most of their countrymen, those who inhabited the swath of territory from just beyond the smoking scar that had been St. Petersburg to Vladivostok, on the rim of the Pacific, believed in their souls they were born to suffer. That made them notoriously stoical. After all, they and their forebears had lived in the heart of the large landmass that the British geographer Halford Mackinder called the world island, so they had endured centuries of attacks from every direction by potentates and panjandrums who coveted their land for its natural riches and strategic position. And they had been afflicted by their own tyrants for centuries: first the czars, then the Communists. But this new calamity was immeasurably worse. However horrible the wars had been, whatever suffering had had to be endured because of the cynicism and cruelty of one despot after another, there had been refuge somewhere. Not this time. This time there was no place to hide.

It quickly grew colder. As was beginning to happen elsewhere, the disappearance of the Sun started a succession of troubles, one bringing on the next, until they collectively changed the definition of what it meant to be alive. The shroud of soot not only brought the cold, it all but ended photosynthesis. Where food production was concerned, it was a double disaster, since agriculture needs both warm air and radiant energy. Two weeks into the darkness, crops were already turning pale and showing a slight brittleness that every farmer in the world knew was a sure sign of the onset of death. So the Russian farmers went out into the cold and harvested as much as they could on the assumption that whatever else lay in store for them, they would at least have sustenance and the means to bring in some rubles.

But what the Russians and their counterparts in much of Asia, Africa, South America, and elsewhere did not know, and what most of Europe and all of North America had good reason to know, was that the food was radioactive. The highest levels were in Southeast Asia, Japan, and Australia, which were immediately downwind from the fifty-seven huge mushroom clouds that had churned over the two belligerents on the subcontinent. But the deadly particles, riding on the jet stream, also swept through the Hawaiian Islands, the Pacific Northwest, and over the Rocky Mountains.

The Rogue and its disintegrated pieces, and then all of the nuclear explosions, also created a temporary electrical field over most of the planet that turned radio and satellite telephone and television transmission into pure static. The atomic and hydrogen explosions radiated electromagnetic pulses, or EMPs, as the physicists called them, that either temporarily shut down or completely fried the brains of digital computers and their power supplies, as well as radios, telephones, Teletypes, and television that used transistors. Since most communication antennae received as well as transmitted, they choked on the surge of electrons that hit them. EMP therefore shut down transportation and communication in many places, some of it permanently. A hundred and sixty-eight airliners lost communication and navigation capability when the Rogue hit and the warheads exploded. Fifty-one were landed safely by pilots and copilots who eyeballed their way to the nearest airports that could handle them and hand-flew their planes onto the runways. The other 117 went down with a loss of 16,554 people.

The communication blackout slowed or stopped altogether news of what had happened, at least during the first two days. But by the third day, word of the multiple catastrophes had spread throughout the United States, Europe, Russia, and elsewhere, some of it over the communication links that still worked, and the rest by word of mouth. Anyone who doubted what he or she had heard had only to look up and see the dark brown clouds that moved continuously from one horizon to the other. The disappearance of the Sun and the consequent drop in temperature was one of the compelling reasons for the farmers to start frantically harvesting their fruit, vegetables, and feed. The president, speaking in his cave and looking as earnest as ever, informed the people of the United States that his worst nightmare had finally come true. Looking hard into the camera, his voice trembling, he explained that large parts of America had been destroyed by the rocks from space, and that the equivalent of a "nuculer" winter was at hand. But America would prevail, he said gravely, because the Good Lord was on its side. Acting either out of bravado or trying to break an air of tension that was palpable in the cave, the president quipped that lower temperatures around the world would have one beneficial effect most people didn't seem to recognize: the cold would reverse global warming. The others in the room smiled weakly or ignored him.

The looting started in Chicago within an hour of the initial impact, as individuals and small groups broke into supermarkets and other food stores in search of canned goods, which they knew were not contaminated. As shelves were emptied by frantic people, many of them cut by the broken window glass, there was more bloodshed as latecomers tried to grab the food from those carrying it. The police did not interfere, and as in the other cities, some either tore off their badges and joined the looters or simply disappeared. Meanwhile the pillaging spread elsewhere, including Detroit, Philadelphia, Denver, Houston, and Los Angeles, where individuals and gangs not only broke into stores, but into homes in Beverly Hills, Pacific Palisades, and throughout the San Fernando Valley, smashing through doors and windows and attacking their frightened inhabitants with any weapon at hand. The Los Angeles Police Department and other law enforcement agencies that still functioned responded by cutting them down on the streets. Rodeo Drive was littered with broken glass, expensive foreign cars with bullet holes in them, and scores of bloody dead and wounded.

Anarchy broke out simultaneously across the Pacific and elsewhere. North Korean peasants and other civilians had endured poverty for decades. But the military, the large standing army from which the family of patriarchal tyrants that led the country drew their power, had traditionally been pampered with the decent food and modern weapons that sapped resources from the others and kept it a threat to the region. The food, decent and otherwise, was now disappearing, together with peasants, who were beginning to starve in rapidly growing numbers. But the guns remained, and so did a nascent nuclear weapons program at Yongbyon that had used highly enriched uranium and Pakistani expertise to produce four atomic warheads. Faced with his own impending end, the plump Great and Esteemed Father who ruled North Korea (and who had an appetite for Swedish blondes) decided to send 6 million men and boys—by far the largest army in Southeast Asia—across the thirty-eighth parallel and into South Korea to raid Seoul's national larder. They were to occupy the country and confiscate everything that could be used to sustain the Father's impoverished and dying domain. And to impress upon the South Koreans that resistance would be futile and destructive, he ordered that one of his four warheads be launched at the port of Inchon a few hours before the invasion.

He counted on the South Koreans and their American puppeteers not knowing that he had expended a quarter of his nuclear arsenal. The Great and Esteemed Father was a vicious sociopath but he was no fool. The idea was to intimidate the South into submission by demonstrating he had nuclear weapons, but not destroy Seoul, which was the prize he sought. Inchon was therefore immolated just before the Great and Esteemed Father's army swarmed across the demilitarized zone that separated the two countries and advanced on the capital. There was little resistance by Republic of Korea forces. The 17,500 American soldiers fought valiantly until they were overrun. The Americans were stationed there to guarantee the nation's sovereignty. Now there was no sovereignty on the Korean peninsula or anywhere else. The warhead that turned most of Inchon to dust further contributed to the darkening cloud that enveloped Earth.

So did the fires. Forest dwellers around the world reacted to the worsening cold by setting large fires in communal areas and elsewhere. Many of them quickly spread to surrounding brush and trees, which were already in the early stages of dying, and which were therefore more combustible than if there had been sun and a clear atmosphere. The forest fires quickly caught up with and consumed every living thing. That included Siberian tigers, snow leopards, many kinds of rare birds, and other creatures that went from endangered to extinct, almost literally in a flash. No one knew about it, of course, but had they known, they would not have cared. Concern over the fate of exotic animals, or any animals for that matter, was an indulgence for those who had the luxuries of being both safe and well-off. Not a soul on Earth was now safe and well-off, and certainly not the inhabitants of the forests themselves, many of whom were burned alive by flames they could not outrun. Uncounted species of insects, birds, animals, fish, flowers, and other flora simply vanished, unmourned, in the dense smoke and flames.

And so to the particulate residue that had been St. Petersburg and other places in Russia, and the deadly radioactive soup that moved with it in the high reaches of the atmosphere, was added thousands of tons of black carbon and soot that rose from the Amazon, Siberia, West Africa, Southeast Asia, what remained of India and Pakistan, Central America, Australia, New Zealand, and other places that were torched so those who set the fires could survive the bitter cold. But of course the fires only made the cold worse. It was a concatenation of terrors, as Sagan had described the effect of the asteroid impact on Yucatán 65 million years earlier.

Old hatreds between displaced Palestinians and Israelis boiled over on the West Bank and in Gaza. It was the same in Rwanda, Congo, and Nigeria, where the Hausa, Yoruba, and Fulani began settling age-old scores with machetes as well as with guns. Whites were reflexively butchered in Zimbabwe, Kenya, and South Africa. The cause of the massacres was not really religious or nationalistic. It was primal. If they were powerless to protect themselves from the wrath of God, as all those people now knew they were, then whatever power they did have rested in their own ability to exercise godly prerogatives. That meant the power to take life. Killing others, the primitives believed in the dark recesses of their minds, affirmed their own lives. Killing validated what remained of their precarious and terrifying existence. To kill was therefore to live.

Earth was now gravely wounded and becoming dysfunctional as one intractable disaster caused another. With oil refineries deserted, the flow of gasoline ended, and so, therefore, did the movement of the highway traffic that carried goods between cities and regions. Worse, with coal mines also abandoned, the combined lack of coal and oil and the desertion of nuclear power stations (many of them dangerously close to going supercritical because of a lack of coolant) brought the generation of electricity to a flickering halt. That ended telephone and telegraph communication and turned home radios, televisions, and computers into useless junk.

Not everyone or everything suffered radioactive contamination from the war on the subcontinent, at least initially. High levels of radioactivity had been anticipated after a nuclear war between the two superpowers. Between July 1945 and November 1962, the United States had set off 216 nuclear explosions to measure their effects in every conceivable situation. Atomic and hydrogen bombs had been set off at the Pacific Proving Grounds in the South Pacific and in the Nevada desert. Every test had had a code name—Hardtack, Fishbowl, Teak, Orange, Starfish Prime, Checkmate, Bluegill, Triple Prime, Kingfish, Argus, Baker, Wahoo, Milrow, Benham, and others—and was designed to measure the damage and other effects caused by the explosion. Some nuclear blasts had been at ground level. Others had been set off in the air. Still others had been underwater at Bikini Atoll and created spectacular eruptions that carried small out-of-commission ships into the air and rolled over and sank larger ones. Those tests, and a great deal of documentary evidence that had been taken from Hiroshima and Nagasaki, eventually went into a large database of information. It, in turn, was condensed by the Department of Defense and the Energy Research and Development Administration into a book, *The Effects of Nuclear Weapons*, which was occasionally updated. A sleeve on the inside back cover even had a handy round plastic "Nuclear Bomb Effects Computer," which looked like a round slide rule and could be used to calculate overpressure, thermal radiation, crater dimensions, initial nuclear radiation, the early fallout dose rate, and other effects of bombs with explosive capacity from the equivalent of a thousand tons of TNT to 20 million tons. The Russians, of course, had conducted their own tests at Semipalatinsk, deep in the southern mountains that border Mongolia, and therefore also had a detailed understanding of the nature and effect of nuclear explosions. It is noteworthy that both sides studied only blast damage from the explosions, not the fires they caused. The actual level of death and destruction caused by the fires could be two to five times greater than that predicted by American planners who did not take them into consideration when they did their calculations.

One of the important lessons learned from all the research was that surfaces contaminated by radioactivity could be washed with water under pressure and air could electronically be scrubbed to remove dangerous radioactive particles. Having long anticipated an interruption of electricity because of electromagnetic pulse or some other effect of nuclear war, both the presidential cave complex in West Virginia and the far larger concrete bomb shelters that ringed the beltway around Moscow had gasoline-powered electric generators that ran air scrubbers in the shelters. They also had stocks of food and beverages (including generous supplies of bourbon in West Virginia and vodka along the beltway) to last two months. But the elaborate facilities merely prolonged the lives of the men and women who hid in them. They could not save them. Into that second month, with no hope of emerging, alcohol consumption increased steadily, as did the use of sedatives administered by doctors.

During the eleventh week, the president of the United States, now always drunk, succumbed to the alcohol, the sedatives, chronically high blood pressure, and a sudden blood clot—deep-vein thrombosis, his personal physician called it—that formed in his right leg and worked its way up his bloodstream and into his lungs in a matter of a few minutes. He died of a pulmonary embolism, turning blue with bulging, bloodshot eyes, as he gasped desperately for air, while in a drunken stupor. The others—the moles, as they had jokingly called themselves almost three months earlier—fared no better. Neither did their Russian and Chinese counterparts. The generators that powered the scrubbers eventually ran out of fuel. And on top of everything else, the level of radioactivity in the privileged sanctuaries crept steadily higher. But it no longer mattered.

There were those on the surface, counted in the many thousands and spread around the planet—mostly near the poles—who somehow either adapted to the radioactivity and toxins or were not seriously affected by them. They found shelter where they could, kept bundled and near fires to stay relatively warm, and fished, hunted, and foraged for sustenance, often cooperatively. Many settled in the partial rubble of their cities and towns and built basic dwellings of brick and wood with their hands. Others did what their distant ancestors had done: they settled in caves. They therefore had some protection from the acid rain. And the stone cocoons held in the heat from the fires. The caves also provided the best environment for bringing children into the dark and poisoned world, and that was done, too.

Before the great catastrophe, before the Rogue came, those who thought about life elsewhere in the universe had considered the possibility that civilizations could become so technologically sophisticated they turned suicidal. If any of those inquisitive individuals had survived the impact and knew about its many consequences, they were left with the knowledge that the inhabitants of their own derelict planet had done precisely that. They had failed to master the means of spotting the Rogue in enough time to stop it, then had worsened the dreadful situation by turning on each other. A small fraction of humanity, most of it scattered and desolate, and some other species, would survive the chain of catastrophes. But the terrible cold, the darkness, and the consequent shortage of food now made the planet deeply inhospitable to the life it still harbored. It was now a derelict world that was partly inhabited by small bands of scavengers, many of them ill, or crazed, or both, who reverted back almost to their race's beginning.

Long before the Rogue and the wars came, scientists had tried to calculate how often a catastrophic, civilization-threatening impact occurs. While the numbers are necessarily vague, the consensus is that such an event happens once every 2 billion years or so. That sounded reassuring enough. But there was an ominous catch. Once every 2 billion years could be tomorrow morning. Meanwhile, one knowledgeable and optimistic scientist had tried to rebut those who warned about an environmental doomsday by saying that Earth was an eminently seaworthy ship to sail on the vast ocean that was called space. And it had been. But it had also been a ship without adequate defenses, insurance, or a lifeboat. It and every creature on it paid dearly for the negligence.

# : 2 :

# LET THERE BE LIGHT

Spaceguard exists. It takes its name from Arthur C. Clarke's novel *Rendezvous with Rama*, which begins with these ominous words: "Sooner or later, it was bound to happen." It did. On the morning of September 11, 2077, he wrote, a thousand-ton rock and metal fireball moving at fifty kilometers a second impacted on the northern Italian plain, destroying in a few moments what it had taken centuries to create. Padua and Verona were wiped off the face of the Earth, and the last glories of Venice sank forever beneath the Adriatic because of the hammer blow from space. Six hundred thousand were killed, and the total damage was more than a trillion dollars. "But the loss to art, to history, to science—to the whole human race for the rest of time—was beyond all computation. It was as if a great war had been fought and lost in a single morning. . . .

"After the initial shock," the master storyteller and visionary continued, "mankind reacted with a determination and a unity that no earlier age could have shown. Such a disaster, it was realized, might not occur again for a thousand years—but it might occur tomorrow. And the next time, the consequences could be even worse. Very well; *there would be no next time.* . . . So began project SPACEGUARD."

Like all compelling science fiction, *Rendezvous with Rama* had a basis in fact. Clarke knew that an "intruder" had exploded over the Tunguska River Basin in the Siberian wilderness on June 30, 1908, scorching and flattening trees and knocking people off their feet forty miles away. He explained at the beginning of his novel that the speeding rock had caused no harm to inhabitants of the area because none were near it when it blew up. But he also made it clear that the line between a harmless impact or atmospheric explosion and a devastating catastrophe in which millions are killed can be frighteningly thin. Given the rotation of Earth, he wrote, Moscow missed being pulverized by a mere three hours and four thousand kilometers. That's the narrowest of margins in the size of this solar system, let alone the larger universe. Neil deGrasse Tyson, an astrophysicist and the director of the Hayden Planetarium in New York, has noted that the scientifically accurate word for impact is *accretion*. But his own preference is *species-killing, ecosystem-destroying event*.

The real Spaceguard is by no means the comprehensive defense system envisioned by Clarke. It was created because scientists began to see conclusive evidence that the possibility of a devastating collision with an asteroid or comet was real. For one thing, it had already happened, as Luis W. Alvarez, his son, and two others showed in a work they published in 1980 that traced the end of the dinosaurs to one or more colossal impacts 65 million years ago. The huge impact crater that was discovered off the Yucatán Peninsula lent credence to the theory, as did the presence of iridium, an element rare on Earth but plentiful on extraterrestrial bodies. So did evidence that an even bigger impact, or a series of them, killed 90 percent of life on the planet 186 million years earlier.

And so, too, did telescopes on Earth that had been trained on the Moon since the time of Galileo and that discovered hundreds of craters, large and small. Some scientists theorized they were caused by volcanoes, while others maintained they were scars left by impacts. (Rock samples and observations by the Apollo astronauts settled the matter once and for all in favor of impacts.) Furthermore, the exploration of the solar system by robotic emissaries from Earth that began when *Mariner 2* scouted Venus in 1962 and *Mariner 4* did the same at Mars two years later also showed that the solar system is a cosmic shooting gallery. Their successors, American and Soviet, carried names such as *Pioneer, Voyager, Luna, Venera, Viking, Zond, Cassini,* and *Galileo.* They sent back pictures showing that every solid body in the solar system, including Mercury, Venus, and the moons of Jupiter and Saturn, were pockmarked with craters, some heavily. The only major impacts ever witnessed by humans happened elsewhere. Over six days in July 1994, awestruck astronomers watched a string of twenty-one fiery fragments of the disintegrated comet Shoemaker-Levy 9 plow into Jupiter's tumultuous atmosphere and disappear in the storms that never end. The blast cloud from each successive impact was larger than Earth.

By the time Jupiter was pummeled by Shoemaker-Levy 9, concerned scientists and engineers in the American Institute of Aeronautics and Astronautics and those in other groups who understood the potential danger posed by the speeding "visitors" had prodded Congress into taking tentative action. In 1990, the legislators authorized NASA to assess the danger from collisions with asteroids and comets and come up with ways to prevent an impact that could end life on Earth. (The technical term for staving off calamity is the decidedly calming *threat mitigation*.) The space agency, in turn, formed two committees. One was supposed to locate what is out there and potentially threatening, and the other was to come up with ways to destroy the threatening visitor or nudge it off course. The detection committee was chaired by David Morrison, an astronomer at NASA's Ames Research Center at Mountain View, California. The interception group was headed by John Rather, NASA's associate director for space technology.

There was no real disagreement about how much damage a given asteroid or comet could cause. The destruction factor could easily be determined by calculating its size, composition, and velocity. The problem—and this continues to fundamentally cripple planetary defense—was not scientific uncertainty. It was about what was realistically achievable given available and projected resources; about reconciling what was scientifically desirable with what was politically possible. It was, in other words, about money. No one in the community doubted that a hundred-meter-diameter asteroid loaded with metal could inflict terrible damage, especially on a densely populated area. And since Earth is struck far more frequently by smaller objects than by larger ones, the chances of a hundred-meter object plowing into the planet are far greater than those of a kilometer-sized one. It is estimated that Earth collides with a 150-meter-or-larger-size rock once every five thousand years, while those in the hundred-meter-and-somewhat-smaller class hit on an average of once a millennium. And there are roughly half a million football-field-size rocks out there. Yet if the worst destruction—generally defined as planetwide and civilization ending as opposed to regional—would be caused by kilometer-or-larger asteroids and comets, then the paltry resources that were to be made available for the search had to be budgeted to reflect that sorry fact.

The committees, or workshops as they were called, issued reports early in 1992 (appropriately, the start of the International Space Year). Morrison's group identified the magnitude of the impact hazard and called for establishing a Spaceguard Survey to begin an accelerated search for potentially dangerous

Earth-crossing asteroids and other Near-Earth Objects, commonly called NEOs, and cataloging their precise orbits around the Sun.

The participants in the workshop concluded that priority had to be given to finding and cataloging Earth-crossers that could destroy most of the world's food crops for a year, and/or result in the deaths of more than a quarter of the world's population, and/or have effects on the global climate similar to those calculated for nuclear winter, and/or threaten the stability and future of modern civilization. "A catastrophe having one or all of these traits would be a horrifying thing, unprecedented in history, with potential implications for generations to come," the report concluded.

Three years later, in 1995, a follow-up study chaired by Eugene Shoemaker (a widely respected planetary scientist and one of the discoverers of the comet that had struck Jupiter the year before) specifically called for an international Spaceguard Survey Program, led by the United States, which would catalog 90 percent of the potentially dangerous Earth-crossers. The astronomers also called for six new telescopes and observatories, plus a Spaceguard Operations Center to coordinate international observations and ensure quick communication between the observatories and astronomers doing follow-up observations.

But the telescopes and observatories were not built, the Space Operations Center was not established, and Spaceguard's annual operating budget was set at a minuscule $3.5 million (though millions of dollars' worth of existing equipment would be enlisted in the program). A pitifully small group of professional astronomers and many amateurs using existing telescopes would have to search for the really big visitors. But they were not mandated to locate and catalog others that could turn San Francisco, Paris, Shanghai, or St. Petersburg into holes filled with smoking and twisted rubble and their inhabitants into particulate matter carried on air currents in the darkened atmosphere. It reflected the political priorities of the time. It still does.

NASA meanwhile correctly took the position that it was the astronomers' responsibility to describe the problem, and if the society for which they did so ignored it, they, at least, knew they had fulfilled their responsibility. It therefore started a Science Definition Team in August 2002, whose task was to study the feasibility of extending the search to potential hazardous objects (PHOs, in the community's jargon) smaller than a kilometer. In late August 2003, the team recommended that 90 percent of PHOs larger than 140 meters be found and cataloged, and that specific ways to accomplish that, including more and better observation systems, be studied. That size was eventually accepted. The twelve-member team noted that current technology is up to the task of finding 90 percent of the smaller rocks within twenty years, and it recommended a combination of ground- and space-based sensing systems, and combinations of them, to do the job. The choice of a space-based telescopic system typifies the kind of realistic compromise that runs through planetary defense and the space program in general. After explaining that some NEOs reflect so little light that they are more easily spotted by heat-sensitive infrared telescopes than by optical ones designed to observe objects in visible light, the Science Definition Team noted that optical systems were not only more readily available than their infrared counterparts, but that infrared systems require "cryogenic" equipment that produces low temperatures. That kind of hardware not only has a relatively short lifetime, but is relatively heavy, which in turn requires larger and more expensive launch vehicles.

And not all danger comes from out there. Earth has an infinite capacity to turn on itself, both naturally and at the hands of the people who live on it. Earthquakes and volcanoes are the products of a dynamic planet whose shifting tectonic plates and surging molten core often lead to massive death and destruction. They are therefore far better known and feared than asteroid impacts, which are abstract to most people, since none have been witnessed by people (though one rock did explode over four Midwestern states on March 27, 2003, sending scores of fragments, one of them five pounds, through roofs). Earthquakes are deadlier and usually do far more damage than volcanoes. An earthquake under Tangshan, China, in 1976 killed a quarter of a million people. And one centered in the Bam region of Iran killed some forty thousand people in early January 2004. Where death and destruction are concerned, however, that volcanic eruption 251 million years ago, possibly in conjunction with one or more impacts, remains in a class by itself.

There is nothing rare about foul weather. Hurricanes, typhoons, monsoons, blizzards, and other nasty and dangerous atmospheric phenomena are a pervasive fact of life. That much hasn't changed in all of this planet's history, and neither has a climate so perverse that it can swing from an age of ice to global warming and then, perhaps, back again. While some dangers on land and sea can be avoided or lessened—forest fires, avalanches, and high seas, for example—people are utterly powerless to deflect or blunt the many threats that are born in the atmosphere. But the dangers can be dodged with enough warning time. The old saw had it that "Everybody talks about the weather, but nobody does anything about it." That expression went the way of the Model T as soon as weather satellites took to space and started providing far more time for weather's potential victims to get out of harm's way.

The race that inhabits this planet has also devised multiple ways to cripple or kill its home and, in the process, itself. They cross the full spectrum from the infinitely subtle to the unimaginably sudden and horrific. Whatever its cause or combination of them, there is now a consensus within the meteorological and wider scientific community that global warming—slow cooking—is under way. Some say the warming of the climate is natural and cyclical. But the usual suspect is carbon dioxide emissions from the burning of fossil fuel, creating a so-called greenhouse effect in which the hot gases pouring out of motor vehicles, airplanes, ships, power plants, and other energy sources that run on petroleum, coal, or natural gas are trapped in the atmosphere the way warm air accumulates in a greenhouse's glass enclosure. And invisible carbon dioxide has now been joined by another culprit, this one all too visible: black soot. NASA scientists reported in December 2003 that dark particulate crud coming from diesel and other kinds of engines, as well as the widespread burning of vegetation in the Amazon, western Canada, the United States, and elsewhere, coats so much of the world's snow mass that instead of reflecting solar rays back into space, which is what pristine white snow and ice do, the soot absorbs the rays. That warms the snow and ice, which makes them melt faster and therefore also increases worldwide warming by as much as 25 percent.

Pollution in its many forms, from atmospheric, to oceanic, to toxic dumps throughout the industrialized world, is well-known because it is easier to see. So is the garbage that is customarily dumped on land and at sea around the world every day. And so, too, is the overkilling of wildlife. Fisheries in both hemispheres are becoming dangerously depleted because a shortsighted and greedy industry is using large factory ships to scour the oceans with huge nets, snaring not only the fish they are after, but the subspecies they feed on, and mammals as well. The industry's drive for short-term financial profit impedes any serious possibility of planning for long-term sustainability. That insidious and perhaps fatal shortcoming—an unwillingness even to try to plan far enough ahead to prevent disaster—runs through all of human enterprise and to the larger issue of the survival of humanity itself. It reflects the imperative of immediate gratification, measured in terms of economic gain that often turns to pure greed, at the price of abandoning long-term environmental security. "It's the economy, stupid," could be Earth's epitaph.

The more dramatic (and noticeable) side of the spectrum includes the steady deterioration of the record of civilization itself, from library books and other paper records that are yellowing and crumbling under the weight of time, to fragile digital systems that are broken into and tampered with, or deliberately destroyed, or else simply evaporate for reasons that are not understood. An article in the journal *Science* noted that 7 million pages of new scientific and medical information are put on the World Wide Web every year and that, unlike hard-copy references (paper), some of that information can change and become "inaccessible." It went on to note that almost 20 percent of Internet addresses in a high school science curriculum became "inactive" between August 2002 and March 2003. So did 108 of 184 Internet addresses for an herbal remedy become inactive within four years. For all practical purposes they evaporated.

The record of humanity's existence in its infinite forms, from cave paintings to totem poles to the Gutenberg Bible to the digitization of thousands of journals in the arts and sciences—its vast cultural achievements, multiple theologies, problematic political history, relationship to the planet and the solar system in which its exists, and a great deal more—is the most precious treasure on Earth because it defines the essence of who we are.

And with the computer becoming the almighty of technology, running and interconnecting communication, transportation, government on all levels, the military, science, medicine, and other fundamental aspects of life around the world, a crash—a massive electronic, and therefore computational, meltdown—would be catastrophic in its own right. The chaotic blackout that struck the eastern United States and parts of Canada in the summer of 2003 because of an electrical grid malfunction in Ohio that spread almost instantly, crippling society in many ways, provided the barest hint of what would happen to an entire continent or to all of Earth itself if the big crash occurs.

Widespread death and destruction by chemical, biological, and nuclear weapons—weapons of mass destruction, as they came to be called as justification for the invasion of Iraq—in regional war or because of acts of terrorism stalks civilization. Nuclear weapons are at the top of the threat list. Barring the sort of miscalculation that could be triggered by one or more asteroid impacts—even relatively small ones—national nuclear war would be so suicidal that starting one does not figure in the strategy of any nuclear power. The use of so-called unconventional weapons by terrorists, however, is a different matter. Those who are eager to kill massively and indiscriminately have two advantages over their enemy: they believe their own deaths will bring eternal spiritual rewards through martyrdom, and their internationalization and nesting in sovereign states makes them elusive and notoriously difficult targets for repayment in kind. Weapons of mass destruction are perfect for those who want to kill massively. That is why superweapon proliferation, and especially the spread of nuclear weapons technology, which is increasing, is so pervasively dangerous.

The world's established political regimes have responded to these multiple threats as their leaders have been conditioned to do: through political expedience. The combination of a short attention span, a marked inability to plan long-term for emergencies, and what has been called a fight-or-flee mentality makes civilization chronically vulnerable to the entire array of disasters. There is some speculation that this social deficit—of concentrating almost exclusively on the most immediate danger—is genetic. While it is true that overcoming immediate threats such as weapons proliferation and terrorism is far more pressing than being able to prevent a long-shot asteroid impact, there does not have to be a choice. The wherewithal to protect Earth should be multidimensional and imaginative. That means outgrowing the tendency to concentrate only on immediate danger and broadening the defense against the wider array of potential threats. This requires resolve and tools. The latter are in place or could be put there. But there is no resolve.

The "conquest" of space (as it was called even before it began) grew out of intense political competition between the United States and the Soviet Union during the Cold War. The plan to colonize a frontier in Earth orbit and then beyond, which was laid out by a succession of visionaries starting with the Russian Konstantin E. Tsiolkovsky in the late nineteenth and early twentieth centuries, and including Wernher von Braun in the 1950s, was abruptly discarded when long-range ballistic missiles were modified to carry military pilots to space for political advantage. That advantage drove both superpowers' manned space programs, and mainly the race to land men on the Moon. It also extended to the robotic exploration of the solar system as far as Neptune.

The race to the Moon, which engaged both superpowers in the mid-1960s, provided a focused goal for their space programs. When it became clear near the end of the decade that the United States would get there first, the Soviet Union refocused on a longer-duration specialty: sending cosmonauts to space for a permanent presence there in orbiting stations. With the Apollo Program successfully completed between 1969 and 1972, with twelve Americans having walked on the lunar surface, NASA started the Space Transportation System. STS, which was sold to the public as "the next logical step," consisted of two interlocking programs: the shuttle and a space station. The idea, first popularized by Wernher von Braun and others in the early 1950s, was to develop a small fleet of reusable space planes that would carry satellites to orbit, help repair defective ones already in orbit, and act as a ferry with which to build the space station.

But STS was haunted by a dire problem: it had, and continues to have, a small public constituency. Apollo had relatively broad support (roughly 50 percent) because it accomplished two objectives that were taken to be important: it landed humans on another world for the first time, which was daring and transcendental, and they were Americans, which seemed to validate the nation and the political system that sent them there. By contrast, the shuttles and the station lacked an explicit, overarching goal. Keeping Americans and their foreign guests in space permanently for reasons that were vaguely articulated, if articulated at all, struck most Americans as irrelevant and wasteful compared to solving the multiple problems they faced on terra firma. What's more, shuttle missions seemed to be so routine until the *Challenger* and then the *Columbia* tragedies that they came to be largely ignored. The Russians had the same problem after the collapse of the Soviet system, with the added headache that came with being broke. Even before the end came, the average Russian, like the average American, had more pressing priorities than keeping their countrymen in orbit. Babushkas who stood on line for hours to buy bread and potatoes were less than enthralled about cosmonauts frolicking in the *Salyut* and *Mir* space stations.

Now the technological tools necessary to reach space and stay there are well understood and at hand. But there is no resolve because there is no apparent need to do so. Yet there is a need. And it is infinitely more important than political and economic competition between nations, and more practical than the amorphous need to explore that has carried science fiction since Lucian of Samosata sent a voyager to the Moon in AD 160. To be sure, humankind is a race whose vanguard has always contained explorers, and the compulsion to learn about new worlds ennobles it. Merton E. Davies, long a mainstay of space reconnaissance and a member of the imaging team that uncovered four planets and their moons on the *Voyager* spacecraft's unprecedented (and unequaled) twelve-year Grand Tour, explained it with characteristic brevity and eloquence: the joy of exploration, he said, "is finding answers for which there are no questions." Further, and this gets to the point, exploring could ultimately turn up places to go if Earth becomes uninhabitable or is destroyed. It is therefore important not only for the intellectual riches it provides but for practical reasons. Yet it is not the most compelling reason to have a space program.

That distinction has to do with the imperative of survival. It has to do with believing in the fundamental sanctity of life in all of its forms and with resolving to protect it by whatever means is necessary. Access to space provides that means. For the first time in Earth's history, its inhabitants are able to defend themselves from celestial and homemade dangers—from the litany of catastrophes described here—because the space age has given them the ability to do so. The overwhelming majority of people are unaware that technology exists that can help humanity survive. Others are too preoccupied with life's daily challenges to care about abstract dangers and how to avoid them.

And still others—numbering in the millions—are convinced that Revelation, at the end of the New Testament, predicts a fiery end of Earth; an apocalypse in which Christ returns and has a climatic battle with the Antichrist. Those who believe in Him are saved. That is, the righteous are destined for heaven, while the others head elsewhere. This is the credo of some religious fundamentalists, much of it trafficked by televangelists on Sunday mornings and by other communicators all the time.

At the height of the Cold War and the flower children's opposition to the combat in Vietnam in 1970, a true believer named Hal Lindsey published this world's epitaph in a book called *The Late Great Planet Earth*. Based on his reading of Revelation and other parts of the Bible, Lindsey predicted that Europe would unite in large part because of fear that the United States lacked the will to resist Communism and would no longer be able to lead the West. He also interpreted Daniel and Ezekiel as foretelling, respectively, the invasion of Israel by an Arab-African confederacy led by Egypt, and by the Russians and the rest of the Warsaw Pact countries, which would finally grab what he called Israel's vast riches. Then the knaves would conquer the Middle East. Lindsey went on to predict the rise of a new and all-powerful Roman Empire led by an absolute tyrant. And in a chapter called "The Yellow Peril," he warned that Red China and other "oriental" nations (sometimes called the Eastern force) would cross a dried Euphrates that bridges the "ancient boundary between east and west" and finally invade the white man's preserve. The prelude to the end of the United States would be internal political chaos caused by student rebellions and Communist subversion, a military that was "neutralized" because there was no one with the courage to use it decisively, and drug addicts winning high political office with support from those dope-crazed, free-loving, anarchistic beatniks and hippies.

The near-end of the world will be exceptionally messy, in Lindsey's opinion, though he explained that it wasn't clear to him whether that would happen by an act of nature or by "some super weapon." He quotes John as predicting that the Chinese alone will wipe out a third of humanity, and that entire islands and mountains will be "blown off the map." This struck Lindsey as indicating an all-out ballistic missile attack, almost undoubtedly by the Chinese Communists, though he quoted Isaiah as predicting, "The earth is utterly broken, the earth is rent asunder, the earth is violently shaken. The earth staggers like a drunken man, it sways like a hut" (verses 19, 20). Isaiah also predicted a curse devouring earth, Lindsey went on, "and its inhabitants suffer for their guilt; therefore the inhabitants of the earth are scorched [burned] and few men are left" (verses 1, 5, 6). Lindsey interpreted these verses and others as predicting "the use of monstrous weapons all over the world." Maybe. But it also could be a kilometer-or-larger asteroid impact. Whatever it is, according to Lindsey, it will signal the battle of Armageddon. And just as the great battle reaches a catastrophic climax, just when it looks as if all life on the planet will be broiled, Christ will make another appearance and save what's left of the world. When that moment comes, Lindsey warned, "are you afraid or looking with hope for deliverance?" Those are very narrow options. There is another.

If the Scriptures are really taken to be prophetic, then Hal Lindsey might profit from going back to God himself, who is famously quoted in the beginning of Genesis as saying, "Let there be light." That could be taken to predict the birth of Thomas A. Edison and the invention of the incandescent bulb. Or, as is generally supposed, as calling for the physical shedding of light on the world, as sunlight. But "light" could also be taken to mean using reason and a rational understanding of nature, as in Enlightenment, rather than adhering to a stunted and atrophied (but profitable) Dark Ages belief system that denies human beings their inherent dignity and instead reduces them to hapless, utterly defenseless victims of their own worst instincts and a violent universe.

History is littered with broken crystal balls, Lindsey's among them. Two decades after he made his dire prediction of an imminent Armageddon caused by global nuclear warfare, a rampaging swarm of millions of yellow people overrunning Christianity and attacking Jerusalem (where God said he would return), and the rise of a European Common Market that is the ten-horned, seven-headed beast described in the book of Daniel, the collapse of the Soviet Union and its European surrogates and the end of the Cold War showed the predictions to be blatant nonsense. Back to the proverbial drawing board. Lindsey, absolutely wrong, but never in doubt, adapted to the times by quietly eliminating the old causes of doomsday and introducing new ones in a book that appeared in 1996. Now the harbingers of the world's imminent demise, also as allegedly predicted in the Bible, are pollution, an environment pushed beyond sustainability by overdevelopment, AIDS, and radical Islamist fundamentalism. (A seer can't be wrong by predicting carnage caused by Muslim terrorists, who have become staples of the crystal ball crowd.)

The fundamentalist mentality, whether Eastern or Western, is undoubtedly what Marx had in mind when he called religion the opium of the masses. And William K. Hartmann, an imaginative planetary scientist with a flair for writing, took on the ultrafundamentalists in a short parody that was published in *Nature* in 2002. It was in the form of a first-person essay, written by a nineteen-year-old applying to a think tank in 2063, about the ultimate triumph of fundamentalism and antienvironmentalist international corporations over a dreadful and misguided humanistic philosophy that taught the scientific method and concepts such as evolution, the plurality of worlds, species extinction, and climate change.

Having described a hopelessly naïve grandfather who thought a golden age was at hand because of mineral extraction in space, the use of solar energy to reduce pollution, and other misguided manifestations of "secular hubris," the smug narrator made his point:

"The problem was that the scientists and academics appealed to so-called facts instead of common sense and faith. They tried to indoctrinate children with godless liberal ideas. . . . They also claimed that humans could build intelligent machines, that life had been created beyond Eden on Mars, and that planets had been discovered around other stars—ideas that would make Earth and humanity merely a random part of nature, instead of the centerpiece of Creation." The fundamentalists ultimately formed a worldwide political system, the enraptured teenager boasted, that ended the funding of traditional scientific research and destroyed many libraries and databases. "That, in turn, led to the end of attempts to control population growth, improve health care, and nurture the environment, and build a human infrastructure in space," he explained, adding that the fundamental reshaping of society gained the Lord's approval. That is why hopelessly naïve throwbacks like his grandfather had to be put away someplace, the brainwashed kid wrote in conclusion.

Whatever else it is, prophesying is lucrative. Michael Drosnin, a former newspaperman, made the bestseller list twice with books purporting to show that the Old Testament does indeed look into the future with profound insight and dread. Drosnin was cleverer than Lindsey. He tried to lend an element of legitimacy to *Bible Code II* with references to Isaac Newton, Carl Sagan, Paul Davies, Francis Crick, and other respected scientists. He also hedged by explaining repeatedly that the multiple catastrophes allegedly foreseen in the Bible were not certain, but only very likely, depending on what is done to avert them. But like Lindsey, Drosnin used the ever-dependable terrorists to buttress his case. "The ultimate danger foreseen in the Bible code," he warned, "is that religious fanatics will get weapons of mass destruction, and make the ancient prophecy real. The countdown has already begun. . . ." This is like saying that eventually someone is going to burn his or her finger. It is bound to come true, but is it prophesy or probability? And being an Ivy League graduate who reads the newspaper, Drosnin knew that Osama bin Laden and his colleagues and successors aren't going to go away. They therefore make ideal villains because they are not only vicious, but permanent. The war against the infidels will continue no matter what happens to bin Laden, Drosin explains the Bible predicts, and Al Qaeda's killers will continue their grisly work after he is gone. Elsewhere he quoted someone named Eliyahu Rips, whom he credits with discovering the Bible code, with having warned Yasir Arafat that the Bible indicated he was very likely marked for assassination. As far as is known, the Palestinian leader died of natural causes in Paris in November 2004, two years after *Bible Code II* was published. Drosnin also credited Rips with predicting that the odds were 100,000 to 1 that there would be an "Atomic Holocaust," a "World War," and the "End of Days" in 2006 based on his reading of the Bible. It was unadulterated, cynical black magic.

As might be expected, scientists and others who tend toward the rational find Drosnin's stuff not only baseless, but harmful in a society that is increasingly obsessed with religious fundamentalism instead of reason. One, Michael Shermer, the author of *Why People Believe Weird Things* and the editor of *Skeptic*, called *Bible Code II* "codified claptrap" in the June 2003 issue of *Scientific American*. The trick, some scientists explained, was to program a computer to find anything the purported prophet wants to find in the Bible or any other book. "When my critics find a message about the assassination of a prime minister in *Moby-Dick*, I'll believe them," Drosnin was quoted as having said. So Brendan McKay, an Australian mathematician, did just that. He found predictions for nine actual political assassinations in the great-whale story, and others in *War and Peace*. Shermer reported that David E. Thomas, a physicist, got his computer to discover that Leo Tolstoy had predicted that the Chicago

Bulls would win the NBA championship in 1998. *Bible Code II* so exasperated Thomas that he put a long rebuttal called "Assassinations Foretold in *Moby-Dick*!" on the Internet. In it, based on how he programmed his computer, he showed that the great-whale story also predicted the assassinations of Indira Gandhi, Leon Trotsky, Martin Luther King Jr., John F. Kennedy, Abraham Lincoln, Princess Diana, and others. Thomas also got his computer to divine this in the Old Testament: "The Bible Code is a silly, dumb, fake, false, evil, nasty, dismal fraud and snake-oil hoax." Shermer has called *Bible Code II* "numerological poppycock" and quoted Niels Bohr, the great Danish physicist, as saying that predictions are difficult, especially about the future.

The good of civilization requires that it continue rather than allow itself to be extinguished, or nearly extinguished, as the small minority that worships death believes has been predicted by the prophets. That means sanctifying and protecting life, literally at all cost, and by any means possible. Sir Martin Rees, a professor at Cambridge University and England's Astronomer Royal, is one of many learned individuals who have eloquently argued for the protection of life on Earth. "Most of us care about the future, not just because of a personal connection with children and grandchildren," he explained in a recent book about the perils civilization faces, "but because all our efforts would be devalued if they were not part of a continuing process, if they did not have consequences that resonated into the far future." One way of preventing the devaluation, he continued, is to send pioneering groups to live elsewhere independently. Rees is convinced that spreading the seed "would offer a safeguard against the worst possible disaster—the foreclosure of intelligent life's future through the extinction of all humankind."

Sending people and their machines to space has in less than a half century created nothing less than a historic revolution in human affairs. While being in space is no panacea—there are formidable problems that cannot be overcome in orbit and beyond—it has realized an immense and diverse potential to both help civilization and protect it from danger. The manned voyages to the Moon over three years, *Voyager*'s twelve-year odyssey, and other monumental missions of exploration have justly been celebrated and appreciated. But as awesome as they were, and as others continue to be, the revolution has taken place incrementally and so quietly as almost not to be noticed.

Residents of the east coast of the United States who are able to follow the progress of hurricanes bearing down on them days before they hit land, and who therefore have time to batten down their homes or flee, are the beneficiaries of the revolution. They have long since come to take long-range hurricane warnings for granted, though those who came before them had far less warning, or none at all. And the very fact they can watch the progress of the hurricanes or blizzards or other threatening weather on television, let alone get news that can affect their lives virtually as it happens, testifies to the quiet revolution in communication. So, of course, does using a small device that weighs ounces to talk with people thousands of miles away. And so, too, does carrying a little machine that tells its user exactly where in the world he or she is. All this and much more is now so integral to human existence, both in the industrialized world and, increasingly, in the less developed one, that it is taken almost as a natural right.

Even less apparent are the robots that circle the planet continuously and routinely monitor natural resources, take readings of crops that help farmers ward off or cure diseases and tell when it is best to harvest, and provide graphic warnings of dangerous pollution and ice breakups that could signal climate change. Others scrutinize Earth for everything from its bulging middle to its core to its precise size to the levels of radiation it gives off. In fact, more has been learned about this planet from space in the last forty years than in all previous history.

Like other revolutions, the one off of Earth has faded into the realm of the normal. The effort to keep people in space by following Von Braun's old script has resulted in the creation of a space station whose existence has never been justified as compelling. That is not to say it doesn't have such a purpose, but only that claiming its primary value is to learn about how the human body reacts to long periods in space, when no missions over long periods are in the works, strikes many as absurd. Worse, the machine made primarily as a ferry to haul space station modules to orbit and get them connected has turned out to be phenomenally complicated and dangerous. It is noteworthy that the destruction of *Columbia* did not cause the national outpouring of grief that the *Challenger* tragedy set off seventeen years earlier. This is partly because the realization that such mishaps can occur was in the public psyche, but probably also because most Americans were in effect looking the other way. It is also noteworthy that seven brave people disintegrated in a $1.8 billion spacecraft whose mission was, in part, to haul spiders, ants, bees, urine, scummy pond water, a magnetized subway-fare card, and other things used in experiments by schoolchildren. There were also many serious science experiments. But the mission as whole clearly reflected what has happened to a space agency that once repeatedly sent its astronauts to another world with a level of managerial skill, technology, and purpose, now faded in history, that were as important as they were inspiring.

But an important and inspiring mission exists. It is to protect Earth and the civilization on it from the dangers of space and to send its seed off the home planet for safekeeping.

# : 3 :

# TARGET EARTH

On June 6, 2002, an intruder similar to the one that exploded over Tunguska in 1908 blew up over the eastern Mediterranean with almost exactly the energy release of the bomb that destroyed Hiroshima. It was tracked by U.S. ballistic missile early-warning satellites whose infrared telescopes registered the flash and instantly reported it to the North American Aerospace Defense Command in Colorado Springs. Obviously, the NEO could have plowed into a populated area and caused massive death and destruction. Less obviously, to get back to Clarke's observation about the Tunguska rock's barely missing Moscow, the one that exploded over the Mediterranean was on the same latitude as the Indian subcontinent.

The explosion occurred as India and Pakistan were rattling nuclear weapons at each other over Kashmir, as usual. But there was an extra element of danger that spring, and it was anything but usual. The Indians were still enraged over a murderous suicide attack on their parliament on December 13 that had left nine Indians and five Pakistani gunmen dead. The Indians were convinced that the audacious massacre was masterminded by Pakistani military intelligence. On top of that, thirty-two people, mostly women and children, were slaughtered and forty-eight wounded in Jammu, Kashmir's capital, on May 14. New Delhi was convinced that Pakistan was encouraging, if not helping, Islamic militants to infiltrate Kashmir and assassinate Hindus. Tensions were so high by May 31—one week before the NEO exploded—that Washington was urging the sixty thousand Americans in India to get out. The prime minister of India, a hard-liner named Atal Bihari Vajpayee, was at that moment considering going to war against the hated Muslims even at the risk of being attacked by their nuclear weapons. Vajpayee later made it clear that he would unhesitatingly have responded to a nuclear attack in kind.

A U.S. intelligence assessment completed at the time of the asteroid explosion estimated that a full-scale nuclear war between the two bitter rivals would kill as many as 12 million immediately and leave up to 7 million injured. "But those are just the immediate casualties," the report said in an apparent reference to millions of others who would suffer the long-term effects of radiation poisoning. Even a limited nuclear exchange would have cataclysmic consequences, the Pentagon report predicted. Hospitals and other medical facilities throughout Southwest Asia and the Middle East would be overwhelmed, and the U.S. military would have to help the victims and assist in the cleanup.

Four months later, when the situation had cooled, the U.S. air force's Brigadier General Simon P. Worden, the deputy director for operations at Space Command (and a trained astronomer), told a House science subcommittee that a tragedy of almost unimaginable proportions could have occurred if the NEO had exploded over India or Pakistan.

"If it had occurred at the same latitude just a few hours earlier, the result on human affairs might have been much worse," Worden told the legislators. "Imagine that the bright flash accompanied by a damaging shock wave had occurred over India or Pakistan. To our knowledge, neither of those nations have the sophisticated sensors that can determine the difference between a natural NEO impact and a nuclear detonation." The United States, by contrast, uses a sophisticated Nuclear Detonation Detection System, which can detect, locate, and report any nuclear detonation aboveground or in near space in near-real time, which is to say almost as it happens. "The resulting panic in the nuclear-armed and hair-triggered opposing forces could have been the spark that ignited a nuclear horror we have avoided for over half a century," the general added.

Nor is it difficult to imagine such a calamity playing out in the Middle East if the day comes when Iran or an Arab nation becomes the second power in the region to acquire nuclear weapons. Israel's early-warning system is far more sophisticated than those on the subcontinent, but a tremendous explosion over its territory—or on it—could nonetheless provoke an immediate retaliatory strike. So could an attack by North Korea on South Korea or Japan bring retaliation by their American protectors.

The exploration of the solar system that began with the *Mariner* mission to Venus in 1962 has yielded an unprecedented windfall of knowledge. Perpetually stormy Jupiter and the retinue of moons that attend it—collectively a miniature solar system—have been studied up close and in remarkable detail. Saturn's hauntingly beautiful rings and their eerie shadows have been imaged and analyzed, top and bottom, by spectrometers that have measured the precise composition of their obediently orbiting rocks and particles. Venus, which is closest to being Earth's twin, is blanketed by a thick layer of carbon dioxide that is hot enough to melt lead and that is therefore the ultimate greenhouse. But the veil was finally penetrated by Soviet robotic landers and then in effect lifted by an American radar explorer named *Magellan*, beginning in 1990. Other emissaries from Earth have gotten good close-ups of Mercury and conducted the Grand Tour that ended at Neptune in 1989. Only Pluto, the distant planetoid, remains unexplored. Earth's own moon, of course, has been studied in great detail by astronomers from Galileo on and was inspected by fourteen astronauts, twelve of them on its surface, from 1969 to 1972.

Part of the cornucopia of knowledge that has been accumulated from the unparalleled adventure that is the exploration of the solar system is this stark fact: the four planets with solid surfaces and the sixty-one known moons that have been scrutinized by telescopes on Earth or up close by visiting spacecraft are all scarred by impact craters, many heavily. Far from being the tranquil place envisioned on starry summer nights, set to music by Gustav Holst, or imagined when taking in Saturn's ethereal grandeur, the universe is abidingly dangerous. Whole galaxies collide with forces that are literally unimaginable.

Asteroids and comets of varying size streak around this solar system at high speed and in all directions all the time. *Viking* orbiter imagery of Mars taken in 1976 showed that the Red Planet is covered with impact craters. Since little asteroids vastly outnumber large ones, most of the craters are relatively small. But some are immense. The Cassini and Huygens craters in the Schiaparelli region, for example, are each nearly five hundred kilometers across, and a third crater, named for the Russian rocket-engineering genius Mikhail Tikhonravov, is almost as large. Others are even larger. They are clear evidence that Mars has taken hits that were in the million megaton range. That's a million million tons of high explosives. Even the planet's two small moons, irregular-shaped chunks of rock and debris named Phobos and tiny Deimos (appropriately named for the Roman gods of fear and dread) carry their own ugly scars. So do Ganymede, Europa, Callisto, Io, and Jupiter's other moons. And the first clear photographs taken of Mercury, returned by *Mariner 10* in 1974, showed a planet so pockmarked by impact craters that it would look like a hideous smallpox victim if it were human.

Shoemaker-Levy 9's assault on Jupiter in July 1994 provided the most vivid and compelling evidence that what could happen to the huge gas ball could happen to Earth, with far more devastating consequences. The comet was discovered, as many other intruders are, by a "mom-and-pop" operation. That is to say, it was spotted by amateurs using a borrowed telescope. It was observed in 1993 by Shoemaker, his wife, Carolyn, and David H. Levy, an amateur astronomer. The trio were searching the night sky for comets and asteroids using a relatively small eighteen-inch telescope at Mount Palomar, the observatory in Southern California.

Light clouds drifted in that night and dulled the ordinarily sparkling sky, so the Shoemakers and Levy debated whether to waste film money trying to take pictures or call it quits. They finally decided to compromise by using partly exposed film. After the pictures had been taken, Carolyn Shoemaker carefully scrutinized them through a stereo microscope. Levy would never forget the moment she spotted the comet. "Suddenly, she sat up straight in her chair and looked more intently," he later recalled. "Then she looked up at us and said, 'I think I've found a squashed comet.'" Levy and her husband each took turns peering into the microscope and confirmed what she had seen: a string of five or six fragments, each with a little tail, boring into the giant planet. The Hubble Space Telescope, which had been sent into Earth orbit three years earlier, would eventually show the number to be twenty-one, though some observers reported seeing twenty-two. Astronomers calculated that the line of fragments—the largest was about three miles long—had been one huge comet that had been torn apart by Jupiter's gravity during a close pass in the summer of 1992. Then it came around again for the final assault.

Hubble's dramatic images of Jupiter under attack as the icy fragments successively slammed into it, leaving a trail of ugly dark spots in its upper atmosphere, galvanized not only the astronomical community, but part of the political one as well. The Chicken Little crowd, which had been studying Earth impacts for years and warning about the danger of another hit, and which had drawn snickers from some other scientists, suddenly gained credibility. "You're going to see this thing take off like a rocket," Reporter George E. Brown Jr., a former physicist who headed the House Science Committee, told *The New York Times.* "It's going to be easy to sell in the Congress," he said about increased funding to spot asteroids and prevent collisions. Drawing an analogy from another field of science, the California Democrat noted that nothing puts federal funding into geophysical research faster than earthquakes that strike major cities. But the analogy was a stretch. Jupiter is far away. More to the point, earthquakes in California and elsewhere happen all the time. As much cannot be said of asteroid impacts, even small ones.

Yet, as noted, the American Institute of Aeronautics and Astronautics and other groups of scientists and engineers who understood the potential danger posed by large speeding rocks and comets had already prodded Congress into taking some action, however tentative. That led to the formation of the two committees, or workshops, that were supposed to devise ways of finding and cataloging potentially dangerous objects and preventing them from impacting Earth.

Funding being finite, hard decisions had to be made about where the real threat started. "Deciding how small NEAs [Near-Earth Asteroids] one should attempt to discover is simply a matter of cost-benefit analysis," Alan W. Harris, an astronomer and senior research scientist at the Space Science Institute in California who specializes in finding errant asteroids and comets, explained. "One must weigh the cost of detection against the 'benefit' in the form of ability to protect against a future impact," he added. David Morrison, who has noted that the cost of locating and cataloging the smaller ones goes up sharply because there are so many of them and they are harder to find, readily agreed.

The scientists in the Spaceguard workshop called for an accelerated effort to locate potentially dangerous Earth-crossing asteroids and other NEOs and precisely catalog their whereabouts. Asteroids in the asteroid belt between Mars and Jupiter and those in the Kuiper Belt beyond Pluto were not taken to be threatening because they orbit the Sun on their own distant tracks and rarely stray off them. But rocks that cross Earth's path, and especially those in the kilometer-and-larger range, were taken to be an altogether different matter. A section of the Spaceguard report titled "Threshold Size for Global Catastrophe" put the problem in perspective:

"The geochemical and paleontological record has demonstrated that one impact (or perhaps several closely spaced impacts) 65 million years ago of a 10–15 km NEO resulted in total extinction of about half the living species of animals and plants. . . . This so-called K-T impact may have exceeded 100 million megatons in explosive energy. Such mass extinctions of species have recurred several times in the past few hundred million years; it has been suggested, although not yet proven, that impacts are responsible for most such extinction events."

The dinosaurs and many other kinds of animals and plants may in fact have been done in by multiple hits occurring almost simultaneously, possibly coinciding with (or causing) massive volcanic eruptions that spewed deadly methane into the atmosphere and the oceans. New research has shown that the Boltysh crater in eastern Ukraine is 65 million years old, and so is another on the floor of the North Sea. Many of the 170 known craters on this planet have never been accurately dated, so the K-T (for the Cretaceous-Tertiary boundary) event may indeed have been caused by a formation of giant rocks, perhaps like the "string of pearls" that slammed into Jupiter. Although the Ukrainian crater is only twenty-four kilometers wide, it is surrounded by a ring of rocky debris that suggests a fiery cataclysm. That kind of impact today would devastate a small country.

"To appreciate the scale of global catastrophe that we have defined," the report continued, "it is important to be clear what it is not. We are talking about a catastrophe far larger than the effects of the great World Wars; it would result from an impact explosion certainly larger than if 100 of the very biggest hydrogen bombs ever tested were detonated at once. On the other hand, we are talking about an explosion far smaller (less than 1 percent of the energy) than the K-T impact 65 million years ago. We mean a catastrophe that would threaten modern civilization, not an apocalypse that would threaten the survival of the human species."

The Spaceguard scientists went on to list the Earth-crossing asteroids one kilometer or larger, with impact energy greater than one hundred thousand megatons, as the greatest hazard, and reported that about two thousand of them were thought to exist, with the whereabouts of fewer than two hundred actually known. (The number has since been reduced to between roughly one thousand and seventeen hundred.)

Next came long-period comets, many of them originating in the Oort Cloud, which is thought to be a sphere more than fifty thousand astronomical units from Earth (an AU is the mean distance from the Sun to Earth, 93 million miles, and is astronomical shorthand for very long distances). The cloud is believed to be composed of comets. Since it is a colossal sphere, comets that leave it and head this way can come from any direction at any time and are therefore impossible to predict and catalog. To make matters worse, they travel at higher speeds than asteroids. Most comets are likened to dirty snowballs spewing dust and gas in long "tails." But their structures are poorly understood. One, called Wild 2, was seen to be a three-mile-wide rock that was pockmarked with craters of its own and had jagged spires when it was discovered early in 2004. And comets break up unpredictably and independently of other objects in space. One long-period comet streaks between Earth and the

Moon on an average of once a century, and one seems to slam into Earth every few hundred thousand years. Warning time of a collision would be a matter of weeks or months, depending on the direction from which it comes. As things now stand, with no protection system in place, that would provide time only for people to find shelter and wait out the devastation.

Finally, the Spaceguard workshop took note of the smaller asteroids and comets, many of which have hit Earth. The one that created the Barringer impact crater near Flagstaff, Arizona, commonly called Meteor Crater, impacted about fifty thousand years ago. It seems to have been mostly iron and hit at a velocity of almost eleven kilometers a second. That means it struck with a force of from ten to twenty megatons and left a bowl-shaped hole that is now a kilometer across and two hundred meters deep.

These "small" rocks were classified as relatively harmless compared to the others. But "relatively" is relative. Asteroids below the one-kilometer minimum that Spaceguard considers capable of inflicting a global catastrophe are nonetheless very dangerous. Those in the two-hundred-meter-to-one-kilometer category could cause a tidal wave so massive it would destroy coastal and some inland cities for many hundreds of miles.

The astronomers estimated that a one-kilometer object strikes with the force of 10 million Hiroshima-type bombs. "These objects constitute the greatest hazard, with their potential for global environmental damage and mass mortality," the survey reported. "Indeed, during our lifetime, there is a small but non-zero chance (very roughly 1 in 10,000) that the Earth will be struck by an object large enough to destroy food crops on a global scale and possibly end civilization as we know it." That object would be the ultimate environmental hazard.

Finally, the members of the workshop called for an international Spaceguard Survey Program, to be led by the United States, whose goal would be to locate and catalog 90 percent of the kilometer-or-larger Earth-crossers. Six special telescopes and observatory buildings, each costing $6 million, were to be built for the survey. The total estimated capital cost of the observatories was set at $50 million, with operating costs of $10 million to $15 million a year. The scientists also called for the creation of a Spaceguard Survey Operations Center, which would coordinate international observations and ensure quick communication between the six observatories and astronomers doing follow-up observations with optical telescopes and radar. The operations center was also to be responsible for computing the intruders' orbits and ephemerides (exactly where they are and will be at any given moment). An asteroid that misses Earth by, say, twelve thousand miles could hit it the next time around, so Earth-crossers that come relatively close are especially worrisome and require close scrutiny. No object in the solar system moves on a path that can be predicted with complete long-term accuracy because all of them are affected by varying forces, including each other's gravitational pull. As noted, however, Congress failed to provide anywhere near the amount of funding that was suggested, so professional astronomers, supplemented by dedicated amateurs, had to make do with what they already had.

Taking their cue from the Americans, astronomers in France, Italy, Germany, Ukraine, the Czech Republic, and elsewhere started the European Near-Earth Asteroid Search Project the same year, 1992, as an informal and loosely structured partnership to supplement Spaceguard. They coalesced into the Spaceguard Foundation four years later.

Understanding that asteroids and comets threaten Earth is far older than Spaceguard. An awareness of the danger can be dated at least to 1301, when Halley's comet terrified people. (It was ultimately depicted by Giotto as the Star of Bethlehem, a sign of hope, in his *Adoration of the Magi*.) An asteroid-Earth collision study was published in 1953. And in 1980, Luis W. Alvarez, his son, Walter, and Frank Asaro and Helen V. Michel published a pioneering work linking trace amounts of iridium in limestone sediments near Gubbio, Italy, with the great K-T extinction event. They concluded that the dinosaurs were done in by a comet ten kilometers across that blasted enough earth into the sky to create months of impenetrable cloud cover. Others believe it was an asteroid. The best-known impact site, the Chicxulub Crater, is on the northwestern Yucatán Peninsula in Mexico. Radar imagery taken from the shuttle orbiter *Endeavour* in February 2000 revealed the barest outline of the crater, which is 180 kilometers across. It is calculated that a crater made by a large impactor is ten to fifteen times wider than the object that made it. If that is true, the monster that made the hole at Yucatán was at least twelve kilometers across. That fact, and countless others, could only be derived from space imagery. The Chicxulub imagery is only part of almost a trillion radar measurements taken from orbit, many of them showing the results of subtle, or violent, events in the planet's history that would otherwise go unnoticed without the long view from space. Using that unique perspective in this and other ways, including for archaeology, meteorology, and resource monitoring, is one of the fundamental benefits of having access to Earth orbit.

The Lunar and Planetary Laboratory at the University of Arizona at Tucson, which has one of the great astronomy departments in the world, began a

Spacewatch Project in 1980 to study asteroids and comets as they relate to the evolution of the solar system itself. Spacewatch uses charge-coupled devices, which are like tiny light meters, to find and track many kinds of objects in real time. The Harvard-Smithsonian Center for Astrophysics in Cambridge, Massachusetts, also tracks and studies large rocks that wander into the neighborhood and is a clearinghouse for observations made around the world. It does what the ill-fated Spaceguard Survey Operations Center was supposed to do.

The creation of Spaceguard and the publication of its highly informative initial report in 1992, followed two years later by the Hubble and *Galileo* imagery and ground-telescope views of Shoemaker-Levy 9's riddling Jupiter, greatly increased awareness of the threat. In 1995, the noted University of Arizona astronomer Tom Gehrels and two collaborators edited a thirteen-hundred-page tome called *Hazards Due to Comets and Asteroids*, a pioneering work that involved more than a hundred contributors who described the nature of the potential danger, its consequences, and means of protection.

In late April 1995, two months after the publication of *Hazards Due to Comets and Asteroids*, a major three-day international conference on Near-Earth Objects, cosponsored by the Explorers Club, was held at United Nations Headquarters in New York. The purpose of the historic meeting was to call attention to the need to collect and interpret information on past Earth impacts, evaluate what was being observed, and set future requirements. The support of astronomical observation programs in both the northern and southern hemispheres—watching all of the space surrounding Earth all the time—was called critically important.

The early and midnineties was a time when the realization of the threat from Near-Earth Objects began to crystallize. Spaceguard came into being, the monumental *Hazards Due to Comets and Asteroids* was published, and the UN meeting was held. Then, starting in December 1995, the U.S. Air Force Ground-Based Electro-Optical Deep Space Surveillance site in Hawaii and the Jet Propulsion Laboratory in Pasadena started a combined operation called NEAT, for the Near-Earth Asteroid Tracking program. JPL is owned by the California Institute of Technology and is therefore NASA's only independent center. It has led the world in solar system exploration since the early 1960s. NEAT uses special software and its own telescopic charge-coupled devices to search the neighborhood for visitors. Within eighteen months of going operational, it had spotted 5,637 new asteroids, fourteen of which were NEOs, five of them Earth-crossers.

The tracking and cataloging program suggested in the Spaceguard report in 1992 finally got under way in 1998 when NASA's Near-Earth Object Program was officially started at JPL and coalesced with NEAT. The goal of the NEO Program's Spaceguard Survey is to catalog by 2008 at least 90 percent of the one-kilometer-and-larger asteroids that cross this planet's path at regular intervals. By October 2002, the survey had found and logged 619, which was impressive. The key to its operation is an automatic impact-monitoring system called Sentry, which was started in the spring of 2002, and which continuously updates the orbits, future close approaches to Earth, and impact probabilities of Near-Earth Asteroids. Its computers use data from the Minor Planets Center at the Harvard-Smithsonian Astrophysics Center in Cambridge, Massachusetts, which in turn routinely receives NEA information from the Spaceguard telescopes. Sentry, which is part of NASA and is also run by JPL, is in constant communication with another impact-monitoring system at the observatory in Pisa, Italy.

Ironically, Sentry's debut in 2002 coincided with the arrival of three new visitors that got the attention of the international news media. This was partly because science writers and other journalists, as well as their editors and producers, were by then aware that accelerated research on the potential danger was going on. And those who crafted fiction were aware of it, too, and thought about its dramatic possibilities. The horrible consequences of run-ins with comets and killer asteroids were dramatically depicted in two Hollywood films, *Armageddon* and *Deep Impact*, that made Earth the target of huge speeding rocks. *Armageddon* is usually derided as sheer rubbish by astronomers, while *Deep Impact* is considered to be remotely credible because its producers initially consulted astronomers, at least superficially, for technical advice.

In keeping with a long tradition that goes back to *The War of the Worlds* and *Dracula*, science was fundamentally distorted to serve the stories' dramatic requirements. Both films drew large audiences and became metaphors for an angry and dangerous Mother Nature. The world was saved in both films by Bruce Willis and Robert Duvall, who blew up the menacing intruders with nuclear weapons only days before they were to do to Earth what a hand grenade would do to a golf cart. In reality, the Willis-Duvall defense would not work, since nuking an asteroid that close to Earth would turn a cannonball into grapeshot: a single large impactor into a formation of smaller ones. But the intention, as usual, was to entertain the audience and maximize ticket sales, and it worked. Donald K. Yeomans, a comet expert who heads the Near-Earth Object Program Office at JPL, and who participated in the original

Spaceguard Survey, is glad *Armageddon* and *Deep Impact* were made, however far-fetched they were. Sure, there was an improbable, saccharine love story involving two teenagers. Sure, the astronomer who spotted the looming menace was killed by a truck driver before he could alert the world. Sure, Willis became an improbable martyr by nuking himself along with his rock. And sure, both endings were absurd. But, Yeomans explained, the movies got people's attention as no scientific tract or meeting of scientists would have.

The public first became aware of potentially threatening Near-Earth Asteroids in 1997, when one called 1997 XF11 briefly made headlines as it seemed to be heading toward Earth. Close tracking and refined calculations eventually showed there was no danger. But XF11 raised public consciousness about the impact threat and figured importantly in starting Sentry. Five years later, three more highly publicized rocks sped through the neighborhood. The first passed on the morning of January 7, 2002. It was named 2001 YB5 for when it was discovered, only thirteen days earlier, by NEAT. YB5 was estimated to be the size of a shopping center (including the parking lots). It missed Earth by 515,000 miles, or a little more than twice the distance to the Moon. David Morrison reported that it was potentially hazardous because its orbit brings it so close to Earth, yet he pointed out that there is no danger of a collision with YB5 for several centuries. But, "if it ever does hit our planet," he added, "an asteroid of this size would excavate a crater (on land) the size of a small city, and the blast would cause considerable damage on the scale of a country like France." The NASA astronomer carefully noted that impacts of that magnitude occur only once in about twenty thousand to thirty thousand years. "However, something as big as YB5 comes as close as this roughly annually," he added. "Thus many thousands of objects come close—a 'near miss'—for every one that actually hits. Most of these are undetected," he continued, "since the coverage of the Spaceguard Survey is limited for asteroids this small."

The allusion to YB5 being small, as dangerous asteroids go, is what makes the subject of asteroid and comet collisions so fearsomely surreal. This potentially menacing object could punch a crater the size of a small city into Earth, or wipe out a city while devastating an area the size of an average nation. Yet the capacity to turn New Bedford, Massachusetts, into a smoking hole, killing more than one hundred thousand people, most of them instantly, only rated YB5 a place in the "mediocre" category on the cosmic-threat chart.

The apparent incongruity of YB5's potential to pulverize an entire city, on the one hand, and the fact that it was not taken to be a high-magnitude menace by astronomers in Spaceguard and elsewhere, on the other, underscores the awesome nature of the potential threat. No scientist would call the destruction of a city trivial. But impact hazards, like the effects of diseases and accidents, are ranked objectively in the order of their potential to destroy. No physician thinks chronic asthma is trivial, for example, but none would rank it as being comparable pathologically to leukemia or pancreatic cancer. So calling YB5 small simply meant that whatever its capacity for destruction, it would not cause a global catastrophe or end life on Earth, which are at the worst-case-scenario end of the cosmic hit parade. And the fact that it passed at a distance of more than a half million miles further diminished its danger. For some scientists the most glaring danger comes not from rocks that pass in the night, but from the journalists who sensationalize the extent of their threat.

The news media function in an environment that fundamentally differs from that of science. They have to inform large numbers of people about important and potentially dangerous events in a way that gets their attention. And they have to do it quickly because news is itself a perishable commodity. The average person has not taken Astronomy 101 and knows next to nothing about science in general, let alone about asteroids and comets. An alarmingly large number of people think that Pluto was named after a dog and that meteorologists study meteors. Furthermore, news organizations in free market economies are financially competitive, so they have to sell their products. Whatever the effect misleading advertising has on selling automobiles or soap, misleading news reporting—the often blurred line between accurate information and entertainment, sometimes called infotainment—can be pernicious. Yet it creeps into the selling of news all the time.

That's why the Cable News Network ran a story that January 7, 2002, captioned, "Earth Escapes Brush with Killer Asteroid." The next day, the *National Post Online* said, "Asteroid Misses Earth by a Cosmic Whisker," while *Florida Today* ran its story under a headline that proclaimed, "Earth Escapes Close Call with Massive Asteroid." The Associated Press sent out a story proclaiming, "Asteroid Big Enough to Raze France Zips by Earth." The story that followed, however, was accurate and anything but hysterical:

"An asteroid large enough to wipe out France hurtled past Earth at a distance of a half million miles just days after scientists spotted it. The asteroid, dubbed 2001 YB5, came within 520,000 miles of Earth on Monday, approximately twice the distance of the moon."

Under a headline that shouted, "Monster from Outer Space Just Missed Us," *The Age* reported, "An asteroid big enough to obliterate a major country missed Earth by a stellar hair's breadth on Monday night, prompting calls from scientists for Australia to join the search for rogue space rocks."

Astronomers and their colleagues in other sciences become disgruntled at such purple prose because "stellar hair's breadth" is irrelevant. If YB5 did not hit Earth, it missed Earth, and whether it missed by one hundred thousand miles or fifty million is beside the point where the result is concerned (unless, as noted, its itinerary calls for ever-closer return visits relatively soon).

On March 8, almost two months to the day after YB5 streaked by, a much smaller asteroid named 2002 EM7 crossed Earth's path at a distance of 174,000 kilometers. This one was a relatively puny rock fifty to a hundred meters in diameter. But it was not seen until four days after it passed because it came from the general direction of the Sun and was therefore not visible in optical telescopes that work at night.

On June 15, it happened again. Another football-field-sized asteroid— which was almost invariably also associated with the Tunguska event— passed at a distance of only forty-seven thousand kilometers. It had been eight years since an asteroid was known to have come that close to Earth, the Near Earth Object Information Center in Great Britain reported. The rock was quickly named 2002 MN. It, too, was not detected until after it had sped by. That MN had snuck up and made its closest approach unnoticed was widely reported with a perceptible snicker. The unarticulated but clear implication in news accounts was that Earth's sole defenders against catastrophe—the astronomers—had been caught dozing. That prompted a pointed rebuttal from Morrison, who deplored what he called "sensationalistic" journalism:

"Some [articles] either decry that the object was found after closest approach (rather than before) or express concern about the 'blind spot' otherwise commonly known to astronomers as the daytime sky. It is quite true that an asteroid close to the Sun in the sky cannot be seen. However, if an NEA is approaching Earth from the daytime sky, it is likely to pass into the night sky, where it can be observed, as this one was. Far from being a cause of concern," the exasperated astronomer added, "the discovery of NEA 2002 MN was another example of the success of the Spaceguard effort in general. . . . Similar concerns were expressed following the discovery of 2002 EMT two months ago, but there is no cause for 'doom and gloom' in either of these asteroids." Yet a sense of apprehension had become palpable, at least among the relatively small, educated minority of laypersons who were interested in the subject, and the journalists who wrote for them.

Spaceguard's growing catalog of NEOs does not make headlines because none have turned up that threaten Earth. Yet one prediction did attract wide media attention when it was made public in April 2002. The news stories were generated from a paper authored by fourteen scientists (including JPL's Don Yeomans) that was published in the respected journal *Science*. Based on careful observations, they wrote, there would be a period of twenty minutes on March 16, 2880, when a kilometer-sized asteroid named 1950 DA would have a "nonnegligible probability"—that is, up to a 1-in-300 chance—of clobbering Earth with a ten-thousand-megaton wallop. The asteroid, as its name says, was discovered in 1950. Then astronomers lost track of it for decades. But it was finally relocated for good on New Year's Eve, 2000. Careful calculations of 1950 DA's trajectory left almost no doubt that it would be heading dangerously close to Earth after fifteen course-altering encounters with the gravity fields of Earth and Mars. "Almost" because of the usual variables, which include how the asteroid's reradiating sunlight would gradually push it onto a new course, the gravitational pull of the other bodies it passes en route here, and the overall dynamic of the larger galaxy. Refining the collision probability—that is, calculating its true course with great accuracy—the authors concluded, might require direct inspection by a spacecraft.

The story of 1950 DA, as the print reporters would have said, made good copy.

The probability of its crashing into Earth was one thousand times greater than the carefully calculated impact probability of any other large object known to be out there. That made it far and away the most potentially dangerous threat to the world ever spotted in space. On the other hand, those targeted for oblivion—if, indeed, they were targeted—had 878 years to think of a way to avoid a catastrophe. Panic was therefore somewhat premature.

One way of deflecting DA, which was discussed when the *Science* article appeared, had to do with something called the Yarkovsky effect. The phenomenon is named after I. O. Yarkovsky, a Russian engineer who noticed that heat from the Sun was absorbed by asteroids and then reradiated back into space, acting like tiny thrusters that pushed the asteroids onto different courses. Joseph N. Spitale of the University of Arizona's Lunar and Planetary Laboratory promptly suggested the possibility of using the Yarkovsky effect to gradually turn DA away from Earth. That would be done by sending a spacecraft out to meet it, joining it in close formation, and then lightening its surface so it would gradually push itself off the collision course with Earth. Spitale made the point that using the Yarkovsky effect would eliminate the need for detonating nuclear weapons. The *New York Times* editorial page picked up on the idea and even mentioned dusting the threatening boulder with soot or powdered chalk or draping it with reflective Mylar.

JPL's Jon Giorgini, the lead author of the *Science* article, spoke for his colleagues when he turned the 1-in-300 odds estimate around. Nothing good could come from an impact, he said with typical understatement. "But a collision between 1950 DA and Earth is so unlikely it is not worth worrying about. And even if it does look like there could be a collision," he added, "we have plenty of time and many ways to deflect the asteroid from its path." Each, including nudging it off course with a laser, would require a lead time measured not in months, or even in years, but in decades.

Not long after 1950 DA was rediscovered, two NASA researchers, Steven N. Ward and Eric Asphaug of the University of California at Santa Cruz, ran a computer simulation of it slamming into the Atlantic sixteen hundred kilometers east of North Carolina at twenty-four thousand kilometers an hour. They found that the asteroid would penetrate to the bottom of the ocean and make a hole three miles deep and blow out a crater almost twelve miles across. It would cause a doughnut-shaped tsunami hundreds of meters high. Within two hours of impact, three-hundred-meter-high waves would crash into the U.S. shoreline from Cape Hatteras to Cape Cod, and within ten hours, twenty-meter-high waves would slam into Europe and West Africa. The effect along the U.S. east coast would be felt four kilometers inland. That would be nowhere near as catastrophic as 1950 DA striking land. But it would definitely not be a good day to be on the beach.

Scientists like Giorgini know they have to balance the possibility of a real threat against poorly substantiated drama—the Chicken Little scenario—where the public is concerned, and that could have been one of the reasons (aside from hard scientific data) he played down the possibility of a collision. As Clark R. Chapman and Daniel D. Durda of the Southwest Research Institute, and Robert E. Gold of the Johns Hopkins University Applied Physics Laboratory, have explained, impact predictions need to be scrupulously checked out before announcements are made to the news media. That way, they have written, astronomers will not be taken to be raising false alarms by an increasingly numbed and indifferent public. The ultimate tragedy, in their view, would be to provide ample warning time that a city-buster was indeed on the way so protective measures, including evacuation, could be taken, only to be ignored by a contemptuously indifferent public.

On July 8, 2003, a handful of concerned and informed individuals who had followed the spate of sightings that had began with Shoemaker-Levy 9 and continued into 2002 and afterward sent an open letter to twenty-four senators and representatives on ten committees about the imperative of addressing the impact threat. Copies were hand-delivered to all two dozen legislators. At least some of those who signed the letter would have gotten the attention of the recipients: Harrison H. Schmitt, a planetary geologist, former U.S. senator, and veteran of Apollo 17, which made the last landing on the Moon; Carolyn S. Shoemaker, who spotted the comet that bore her name, and then of the Lowell Observatory; Neil deGrasse Tyson, an animated astrophysicist and the director of the Hayden Planetarium in New York, who is in perpetual overdrive; Freeman Dyson, then president of the Space Studies Institute in Princeton; Thomas D. Jones, a planetary scientist with a doctorate and a former astronaut on four shuttle missions; and John S. Lewis, a widely respected professor of planetary sciences at the University of Arizona.

"We write to you today as concerned citizens, convinced that the time has come for our nation to address comprehensively the impact threat from asteroids and comets," they began, making it clear that none of them was an "alarmist.""A growing body of scientific evidence shows that some of these celestial bodies, also known as Near Earth Objects (NEOs), pose a potentially devastating threat of collision with Earth, capable of causing widespread destruction and loss of life. The largest such impacts can not only threaten the survival of our nation, but even that of civilization itself." The letter suggested three courses of action that warranted federal support:

1. NEO Detection: Expand and enhance this nation's capability to detect and determine the orbits and physical characteristics of NEOs.
2. NEO Exploration: Expand robotic exploration of asteroids and Earth-approaching comets. Obtain crucial follow-up information on NEOs (required to develop an effective deflection capability) by directing that U.S. astronauts again leave low-Earth orbit . . . *this time to protect life on Earth* (italics added).
3. NEO Contingency Planning: Initiate comprehensive contingency planning for deflecting any NEO found to pose a potential threat to Earth. In parallel, plan to meet the disaster relief needs created by an impending or actual NEO impact. U.S. government/private-sector planning should invite international cooperation in addressing the problems of NEO detection, potential hazards, and actual impacts.

"For the first time in human history, we have the potential to protect ourselves from a catastrophe of truly cosmic proportions. All of us remember vividly the effect on our nation of terrorist strikes using subsonic aircraft turned into flying bombs. . . . Consider the ramifications of an impact from a relatively small NEO: more than a million times more massive than an aircraft, and traveling at more than thirty times the speed of sound," the letter continued. "If such an object were to strike a city like New York, millions would die. In addition to the staggering loss of life, the effects on the national and global economy would be devastating. Recovery would take decades." Then a warning: "If we do not act now, and we subsequently learn too late of an impending collision against which we cannot defend, it will not matter who should have moved to prevent the catastrophe . . . only that they failed to do so when they had the opportunity to prevent it." The lawmakers did not respond.

Earth also has an infinite capacity to turn on itself. Those who have experienced the effects of earthquakes and volcanoes, as well as many other destructive forces of nature, including ferocious weather, are neither contemptuous of warnings nor indifferent to the possibility of imminent death and devastation.

Earthquakes and volcanoes are the products of a dynamic planet and are better known than space rocks because they strike far more frequently. (As far as is known, no one has ever been done in by an asteroid or comet, though there have been some close calls, as when an asteroid exploded over four Midwestern states on March 27, 2003, sending scores of rocks through roofs. Another one, weighing almost three pounds, crashed through the roof of a home in Auckland, New Zealand, on June 12, 2004. "It was like a bomb had gone off," said the lady of the house, who was in another room at the time.) Each is pernicious in its own way. With at least two notable exceptions, earthquakes are deadlier than volcanoes and do far more damage. But the exceptions, if a growing number of paleontologists and other scientists have it right, caused catastrophic deaths on a global scale.

The first is believed to have been that eruption 251 million years ago that caused the Permian-Triassic transition. The second seems to have been the intensive period of volcanic eruptions that, together with one or more impacts, are thought by most interested scientists to have killed half of the life on Earth, including the dinosaurs, over a period of time. Earthquakes along the San Andreas Fault in California are a well-known scourge. Spot shocks, such as the one that struck the town of San Giuliano di Puglia, in southern Italy, on

October 31, 2002, killing twenty-six children and a teacher in a nursery and elementary school, and two other adults in another building, are relatively common in the many earthquake zones around the world. On May Day, 2003, another struck outside Bingöl, in eastern Turkey, killing more than 150 people, many of them also schoolchildren. But recalling an earthquake along the North Anatolian fault that had killed seventeen thousand Turks four years earlier, one knowledgeable geologist called the Bingöl quake "a miss." Three weeks later, the region east of Algiers took a direct hit, when yet another one struck and flattened Thenia, Boumerdès, Dellys, Bordj Menaiel, and other towns along a fifty-mile stretch near the Mediterranean, killing more than twenty-two hundred and injuring at least ten thousand others. The unfortunate residents of the area live on a major fault line that stretches from Portugal to Sicily. And much more recently, as previously mentioned, an estimated forty thousand hapless souls—more than a third of the population—were crushed to death in Bam, Iran, when a quake struck there on December 26, 2003. Within ten days, the head of Iran's Supreme National Security Council announced that it was thinking about moving the capital out of Tehran, a city of more than 12 million that is in one of the most earthquake-prone regions in the country.

A year later to the day—December 26, 2004—another monster quake struck, this time just to the west of the northern tip of Sumatra at a minute before eight in the morning. That eruption was especially deadly because it happened underwater and therefore started a tsunami that fanned out across Southeast Asia and even reached the shores of East Africa. Tsunamis are rolling surges of water that are started by undersea landslides, volcanic eruptions, or asteroid impacts. Two and a half hours after the eruption, a wall of water as high as twenty feet slammed into India and Sri Lanka, smashing and drowning an estimated forty-one thousand people. It also struck nine other countries, including Malaysia, Thailand, Somalia, and other parts of Indonesia, where one hundred thousand others perished. With so many washed out to sea, accurate numbers of people lost were hard to come by, but it was estimated to be more than 150,000. To make matters worse, many thousands of survivors were left without food, shelter, or drinking water and had to live in unsanitary conditions that invited cholera, malaria, and other diseases. It was an international disaster of historic proportion.

Secretary of State Colin L. Powell, who took in the devastation from a helicopter as well as on the ground, said that he had been through wars, hurricanes, tornadoes, and other times of sheer violence, but had never seen devastation to equal that left by the tsunami. Sri Lanka's civil war killed sixty-four thousand people in eighteen years. The tsunami killed half that many in less than an hour.

Three days later, Clarke, who had vacation bungalows at two beachfront locations on Sri Lanka, told Benny Peiser that his family and retainers had escaped harm, but that others were less fortunate. Peiser synthesizes climate news on the Internet. It was a disaster of catastrophic proportions, Clarke said. There were more than two million natives and many foreign tourists on Sri Lanka for the Christmas holiday, which suddenly turned into a horrendous nightmare. After noting that the nation had no real resources to cope with the devastation, Clarke called for it to upgrade its technical and communication facilities so casualties could be reduced in disasters.

A network of tide gauges and seismographs, which is strung out in the Pacific to warn of tsunamis there, could be strung out in the Southwest Pacific and the Indian Ocean. And spacecraft have their place as well. Satellite imagery of the area quickly established the extent of destruction by showing, for example, Banda Aceh on Sumatra intact before the tsunami and cut in half by flooding afterward. Studying the pictures could help scientists learn why some areas were harder hit than others, which would help in planning for future disasters. Other imagery showed that Myanmar, the region's most politically isolated country, did not seem to have suffered much at all. And there is another way spacecraft can help in such emergencies. NOAA weather satellites that are modified to pick up such large disturbances from near-polar orbit could provide warning time for evacuation, even if only an hour. But that, in turn, would require an international communication system that could get the warning to vulnerable areas quickly. Isolated coastal villages would have to be connected to the warning system so word of the approaching tsunami or other mass killer would immediately be received. That in itself would be a formidable task.

About fifty volcanoes erupt every year, and two or three eruptions a decade cause multiple deaths and severe destruction. The magma, or molten rock, they bring to the surface, sometimes snaking down the volcanoes' sides in glowing rivers, or else blasting high into the air with thunderous explosions, is the fiery essence of Earth itself. Lava carries the heat that formed the planet.

Volcanoes' role in the assault on Earth is not so much about death and destruction, though there has been plenty of both, as assaults on the environment. Unlike earthquakes, which basically involve violent shaking of the ground and often the opening of chasms, volcanoes are statuesque; some might say majestic. Vesuvius, which buried Pompeii, Herculaneum, and Stabiae under mounds of cinders, ashes, and mud nearly two millennia ago, and Krakatoa, in what is now Indonesia, are the stuff of legend. Krakatoa set off a series of earthquakes in May 1883, followed on August 27 by one of the most awesome explosions in recorded history. The volcano, which was in the Sunda Strait, between Sumatra and Java, literally blew itself and the island it was on out of existence. Six cubic miles of rock and ash were flung twenty miles into the sky, chunks of glowing debris rained down over hundreds of square miles, and tsunamis rolled out in all directions, sinking ships and boats as far away as Calcutta, two thousand miles away. Krakatoa claimed more than thirty-six thousand victims and left a thousand-foot-deep hole in the ocean floor. Many thoroughly terrified people, seeing the lingering, dark ash cloud created by the debris, and watching bizarre red, blue, and green sunsets, became convinced that the end of the world was at hand. And that meant the dawn of Judgment Day.

But Krakatoa wasn't even a contender for that distinction. The biggest subterranean explosion in recorded history happened sixty-eight years earlier inside an obscure thirteen-thousand-foot-high volcano, also in Indonesia, called Tambora. That one began rumbling in early April 1815 and blew its top on the fifteenth, blasting twelve cubic miles of gases, dust, and molten rocks into the atmosphere and onto its island, Sumbawa, where ten thousand died immediately. Many of them were overtaken by the lava, which poured down the slopes at an estimated hundred miles an hour, burning grasslands and forests along the way. A thick plume of gas and smoke climbed twenty-five miles high.

But it was the 200 million tons of dust, sulfurous gas, and rocks, earth, and particles of all sizes that Tambora belched into the sky, and which mixed with water vapor, that attacked the whole planet. The thick, dark shroud soon blocked sunlight across the northern hemisphere. That turned summer into winter. Unseasonably cold weather in China and Tibet killed trees, rice, and water buffalo. The deep chill also killed another ninety thousand inhabitants of Sumbawa itself and the surrounding region. Since huge tracts of crops were devastated, most who managed to survive the explosion starved to death. Even a year later, in mid-May 1816, the weather in the northeastern United States turned "backward," as some Yankees put it. Summer frost struck the East Coast from Virginia to New England. Pharaoh Chesney, a Virginian, would later recall sleigh riding in June and seeing snow falling on Independence Day. Thomas Jefferson, retired in Monticello after two terms as president, had such a poor corn crop that year he had to apply for a $1,000 loan. It was much the same across Europe, where farmers were severely punished by a cold spell that wouldn't quit.

If there was a bright spot in the persistent gloom, it was cultural. Lord Byron, Percy Bysshe Shelley, Shelley's fiancée, Mary Wollstonecraft, and John Polidori were vacationing near Lake Geneva that dark summer. They entertained each other by reading German ghost stories during one storm in June and, with their imaginations captivated, decided to pen their own Gothic tales. Byron captured the foreboding mood in his poem "Darkness," in which he wrote, "The bright sun was extinguish'd" and "Morn came and went—and came, and brought no day." Polidori wrote *The Vampyre*. And the future Mary Shelley was inspired by the unrelentingly dismal and depressing weather to begin writing *Frankenstein*, which to this day remains a classic morality lesson about the consequences of tampering with nature.

The danger from celestial visitors, nasty weather, earthquakes, and major volcanoes occasionally blowing their tops is relatively slight compared to what people can inflict on each other with assorted weapons. These come in two generic versions: those used against large numbers of people and those used against Earth itself. They were called superweapons, or unconventional weapons, during the Cold War. Now they are called weapons of mass destruction, or WMD, and they come in three versions: nuclear, chemical, and biological.

Those who created the atomic bomb in the United States during World War II, as well as their Soviet counterparts and the Chinese and others who came after them, invented a menace of unprecedented proportion. And the chances that their creations will destroy large tracts of Earth far surpass the equivalent danger from nature. An estimated seventy thousand Japanese were incinerated instantly or died from their wounds within a month of the attack on Hiroshima. And seventy thousand of the city's seventy-six thousand buildings were severely damaged or destroyed altogether. "It is no exaggeration to say that the whole city was ruined instantaneously," one Japanese study reported. Nagasaki fared slightly better because its steep hills confined the explosion, though the deaths, injuries, and destruction were indescribable. Richard Rhodes, whose *The Making of the Atomic Bomb* chronicled both attacks, reported that those who suffered radiation poisoning at Nagasaki spoke with "eloquence of unspeakable suffering."

The horrendous level of death and destruction that came with the atomic bombing of both cities would have paled in comparison to the results of a nuclear war between the East and the West in the decades that followed. The two relatively puny bombs dropped on Japan had grown to many hundreds of more powerful ones on both sides by the time the Cuban missile crisis erupted in October 1962. By then, the concept of mutual assured destruction, commonly called MAD, had taken hold. It rested on the premise that a nuclear attack by one side against the other would be answered in kind and would therefore amount to national suicide.

"The proponents of assured destruction like to use the euphemism 'unacceptable damage' in order to avoid explicit mention of poisoned earth and burned bodies," a disgusted Freeman Dyson wrote in *Weapons and Hope*, which was published in 1984, at the height of the Cold War. "They calculate in various ways the number of cities which have to be demolished and the number of people who have to be killed to do 'unacceptable damage to Soviet society,'" he continued. "Unacceptable damage seems generally to require a few tens of millions of corpses."

Martin Rees, another outstanding physicist, is deeply pessimistic about this planet's future because of technology that can lurch dangerously out of control and an overstressed environment's inability to sustain life. He has written that the body count mentioned by Dyson came closest to happening during the crisis over the Soviet missiles in Cuba and has quoted two of President John F. Kennedy's closest aides, both of whom were intimately involved in the crisis, as believing it as well. "This was not only the most dangerous moment in the Cold War," said Arthur M. Schlesinger Jr., a historian and special assistant to JFK, at a conference marking the fortieth anniversary of the crisis. "It was the most dangerous moment in human history. Never before had two contending powers possessed between them the technical capacity to blow up the world." Robert S. McNamara, who was secretary of defense at the time of the standoff and who came up with the concept of mutual assured destruction, was at the meeting and agreed. "I believe that was the best-managed Cold War crisis of any, but we came within a hairbreadth of nuclear war without realizing it," he said. "It became very clear to me as a result of the Cuban missile crisis that the indefinite combination of human fallibility (which we can never get rid of ) and nuclear weapons carries the very high probability of the destruction of nations."

The late McGeorge Bundy, who was Kennedy's special assistant for national security during the crisis, was more sanguine. He wrote that both Kennedy and Khrushchev were determined to avoid what was then called "the unthinkable" and were in control of the situation. "Even at the climax of the crisis," Bundy wrote in *Danger and Survival*, a richly detailed history of the first half century of nuclear weaponry, "Kennedy was worried not about the immediate resort to nuclear weapons, or even about the direct impact of a first or second step, but rather about the risks that might arise as one step followed another." Whatever the truth of the matter, and it is very interpretive, there is no question but that the specter of unintended consequences stalked both sides during those tense thirteen days.

Twenty years later the United States alone had roughly twenty-six thousand nuclear weapons and an annual budget of more than $35 billion to care for them and invent new ones. The Soviet Union, never as well off as its capitalist antagonist, had something like ten thousand fewer weapons and a lower budget. But everyone knew that was more than enough.

The heart of U.S. nuclear war-fighting doctrine in the formative years of Dwight D. Eisenhower's presidency, between 1952 and 1960, rested on a massive, preemptive strike and huge overkill. In 1960, separate air force and navy nuclear-target lists that totaled about twenty thousand places earmarked for obliteration were combined into a fully coordinated Single Integrated Operating Plan. The new integrated scenario, or SIOP, called for attacking Soviet cities the size of Hiroshima with four nuclear weapons, three of which had an explosive capacity of a megaton each, or the equivalent of a million tons of high explosive, while the fourth bomb was to be a four-megatoner. Seven million tons of destructive energy would be used on a target that, in the case of Hiroshima, had been devastated by thirteen thousand tons. It would have been the essence of "overkill."

The notion that nuclear war would almost undoubtedly destroy all life on Earth—the ultimate act of mass suicide—was accepted by every leader on both sides. Dwight D. Eisenhower, a career soldier, was deeply frightened by nuclear weapons. At a National Security Council meeting on August 4, 1955, whose main subjects were war gaming, target selection, and civil defense, Secretary of Defense Charles E. Wilson raised the possibility of the Soviet Union sending a colossally powerful H-bomb into an American harbor on a ship or submarine.

That prompted Ike to think about the worst-possible catastrophe. "How much force would it take to knock Earth off its axis?" he asked idly. A Department of Defense scientist told him that the matter was under study. A powerful hydrogen bomb exploding offshore could cause a two-hundred-foot-high tidal wave that might actually start the whole planet wobbling, he said. "We finally will get destruction of such magnitude that you can't talk about defense," the president told the twenty-five men sitting around the table. And how much would it take to make Earth radioactive? he wanted to know. A thousand megatons would be close to the tolerance point, someone else answered, and ten thousand megatons exploding in the atmosphere would decisively poison the whole world. Advances in nuclear weapons technology had come so quickly, Eisenhower mused, that both superpowers would "soon get to the point where no one can win." They were already there.

Although it is never articulated, every government that contends with an enemy government dreads the prospect of its citizens in effect walking away from the conflict; of deciding that the reasons for potential war are unjustified and the outcome is unacceptable. Martin Rees indirectly raised that possibility when he discussed the odds of going into nuclear war and whether, ultimately, the carnage is worth it. "I personally would not have chosen to risk a one in six chance of a disaster that would have killed hundreds of millions and shattered the physical fabric of all our cities, even if the alternative was a certainty of a Soviet takeover of Western Europe," he has written. Faced with the ultimate act of destruction, he believes, life takes precedence over politics, however oppressive the system. That attitude—let'em have it; the game isn't worth the candle—is of course anathema to every government on the planet.

Both superpowers therefore relied on four traditional strategies for preventing such a scenario: arousing patriotism, vilifying the enemy, maintaining that a war could be won, and holding out the possibility that ordinary citizens could not only help stop a nuclear air attack, but that careful preparation by individuals and groups would ensure survival even after enemy bombers, and later missiles, got through. However effective that last one may have been against the Luftwaffe's and the RAF's thousand-pounders, it was patently absurd where atomic, and then hydrogen, bombs were concerned.

People were told they were protected by a civil defense system that held the hope of survival even after an attack. That was extraordinarily cynical. Not only were the effects of the attacks on Hiroshima and Nagasaki well understood, but so were the extensively documented effects of 1,054 nuclear tests that were conducted in Nevada and the South Pacific between July 1945 and September 1992.

The extraordinary damages, structural and biological, caused by the carefully monitored tests were kept hidden from the public. But the tests were used to prepare an *Emergency Plans Book*, which amounted to a doomsday scenario that described the near-complete devastation that would be caused by a thermonuclear attack on the United States. It was first published by the Department of the Air Force in 1958 and classified secret. Then it was accidentally declassified in 1998, and when the error was discovered, it was hastily reclassified a year later.

In fostering the notion that the residents of target cities had a hope of surviving a nuclear explosion by hiding in basement shelters, the Pentagon was grossly distorting well-understood facts to the contrary. The *Emergency Plans Book* predicted that almost one in five Americans would perish in an attack and millions of others who were injured by debris or poisoned by radiation would quickly exhaust emergency medical facilities. Ninety percent of the hospital beds in the country would be destroyed. Burn victims, a medical specialty, would immediately overwhelm the physicians and facilities for treating them. (Years after the book was written, a group called the Physicians for a Responsible World confirmed it and warned that a single nuclear bomb dropped on New York would create more burn victims than all of the specialized burn beds in the country could handle.)

That, of course, still applies. And so does the potential for social chaos, which the *Emergency Plans Book* also described. Survivors who worked their way out of the shelters would be confronted by dysfunctional economic and governmental systems. "The attack has caused an almost complete paralysis in the functioning of the economic system in all of its aspects," the book explained matter-of-factly in the present tense. "There is an immediate severe impact on organized governmental activities, a fragmentation of society into local groups, a deterioration of our social standards, a breakdown in our system of exchange, and complete disruption of normal production processes." That kind of social deterioration hinted darkly at a breakdown of law and order and violent conflict over scarce resources such as food, clothing, and medical supplies. And although the fate of America's art treasures and historical artifacts was not specifically mentioned, they would also remain at grave risk. The National Archives has a special vault for the Bill of Rights and the Declaration of Independence, while the National Gallery of Art and other major museums have evacuation routes out of Washington and into rural Virginia for priceless art treasures. This presumes there would be sufficient warning time to rescue them. But experience indicates otherwise. Warning time in the event of what Rees calls a successful nuclear megaterrorist attack using a bought or homemade weapon, as indicated by the attacks against the World Trade Center and the Pentagon, would be zero. And as the wholesale looting of treasures from Baghdad's National Museum during the war there in the spring of 2003 showed, art and antiques are not safe from greedy citizens, either.

The opposing governments on both sides of the Iron Curtain carefully kept secret the ghastly result of an all-out war for fear of frightening the citizens who might have to fight it and suffer the consequences. But writers, scholars, songwriters, comedians, and filmmakers felt no such compunction. They described the "unthinkable," sometimes as cautionary tales. In one of the first, a prolific novelist named Nevil Shute envisioned a world that was ended, not by incineration from sudden, massive nuclear explosions, but by the insidious and relentless effects of unseen deadly radiation that drifted around the world after an accidental nuclear war. He described the ultimate tragedy in *On the Beach*, which was published in 1957, and which was made into a film. The book's title was an apparent metaphor for a place where living things wash up and die. The story ended mournfully when a young woman bade good-bye to her American lover—a naval officer whose submarine slowly disappeared into the consuming mist—and called after him about reuniting in heaven. "Dwight," she said, "if you're on your way already, wait for me." Then she washed down poison with brandy to hasten the inevitable.

Novels like *On the Beach* (only rabbits survived), *Seven Days in May*, *Fail-Safe*, and *Red Alert*, and certainly the popular films that were made from them (*Red Alert* became *Dr. Strangelove*), went far in raising the consciousness of Americans and others about the real extent of the nuclear peril at a time when both sides were increasing their weapons stockpiles and the number and variety of ways to get them to their targets.

So did a number of nonfiction books. *Missile Envy: The Arms Race and Nuclear War* was written by Dr. Helen Caldicott, an Australian pediatrician, and published in 1984. It took the United States squarely to task for the dangerous state of the world by using political moves selectively. The book's title was a play on another form of envy that is well-known in psychology. *Nuclear Ambitions*, written by Leonard S. Spector with Jacqueline R. Smith under the auspices of the Carnegie Endowment for International Peace, traced the proliferation of nuclear weapons throughout the developing world in painstaking detail. Janne E. Nolan did the same for ballistic missiles in *Trappings of Power*, which was published by the Brookings Institution in 1991. *On Thermonuclear War*, written by the Rand Corporation's insightful Herman Kahn, was a lengthy, densely packed, and even-handed treatise on nuclear-war scenarios, their effects, and ways to deter them. It was widely read when it appeared in

1978. His *Thinking about the Unthinkable*, which came out in the 1980s, was a much shorter follow-on. And if the likely consequences of a nuclear war were still not clear, Jonathan Schell's classic, *The Fate of the Earth*, accomplished that with meticulous reporting and eloquent prose. Schell was deeply wary of human nature and the flirtation with megaweapons. "As we built higher and higher," he wrote, "the evolutionary foundation beneath our feet became more and more shaky, and now, in spite of all we have learned and achieved—or, rather, because of it—we hold this entire terrestrial creation hostage to nuclear destruction, threatening to hurl it back into the inanimate darkness from which it came."

But it wasn't all Sturm und Drang. Comedians and balladeers like Tom Lehrer, a Harvard mathematics instructor whose songs of social satire turned him into a folk hero in the sixties, also warned about the menace of superweapons. Turning to war nostalgia, for example, he said, "I feel that if any songs are gonna come out of World War III, we'd better start writing them now. I have one here. Might call it a bit of pre-nostalgia." That introduced "So Long, Mom," a song sung by an airman as he carried nuclear bombs to war. In another, "The Wild West Is Where I Want to Be," he described a West transformed from the cowboy era to one large test range for missiles and atomic weapons. In a well-known reference to taking precautions to protect oneself from radioactivity, he explained that he would wear a pair of Levi's "over my lead BVDs." (That stood for Bradley, Voorhees & Day, a popular brand of men's underwear that was bought by Fruit of the Loom.) And like the comedian Bob Newhart, Lehrer wrote a stinging parody of Wernher von Braun, the German rocketeer and visionary who sold his services to the United States after World War II and was instrumental in getting Americans to the Moon.

The final ten years of the twentieth century began optimistically enough. The specter of Earth being turned into a nuclear furnace because of some act of Strangelovian madness began to fade with the end of the Cold War as the old rivals, one of them an economic cripple, backed off from round-the-clock alert and agreed to reduce the number of strategic nuclear weapons. The world seemed to be safer than at any time since the beginning of the nuclear age. But there were two reasons why that was not so.

The profound differences between the two superpowers in the Cold War—essentially more nationalistic than ideological—and the larger political arena in which East-West competition functioned, dominated and stabilized international politics. The two antagonists, in other words, had kept their respective alliances, NATO and the Warsaw Pact, on tight leashes and effectively did the same in most—but not all—of the rest of the world. But as the old framework that kept order started to erode, some suppressed developing nations and large transnational groups with grudges that went back to nineteenth-century colonialism began to try to fill the vacuum. And they gave early signs of intending to wrest their destinies from the industrialized, Western-oriented world with force. As R. James Woolsey, the director of the CIA during the Clinton administration, said at his congressional confirmation hearing in February 1993, roughly three years after the USSR imploded, "Yes, we have slain a large dragon. But we live now in a jungle filled with a bewildering variety of poisonous snakes."

The snakes were driven not only by long-standing political and economic grievances, but by implacable envy and cultural hatred. The second reason the world was not safer, and in fact was becoming even more dangerous, was that the snakes understood that fighting the industrialized nations on their own terms was futile. They therefore resolved to use unconventional warfare—then called guerrilla warfare or insurgency, and now called terrorism—and do so with the ultimate equalizers: the weapons of mass destruction.

The weapons themselves had been proliferating long before the end of the Cold War. China tested its first atomic bomb in 1964, driving India, its nemesis, to follow suit ten years later. New Delhi claimed, straight-faced, that the purpose of its underground nuclear explosion in 1974 was to test a method of moving large amounts of earth. The Pakistanis, fearing that the earth India wanted to move was in their country, began their own nascent nuclear weapons program with the help of India's other regional rival, China. There were no more Indian tests until 1998, when the newly elected Hindu nationalist party ordered five test explosions. Pakistan reciprocated two weeks later. "Today," said Pakistan's prime minister in announcing the test, "we have settled the score."

But the old religious enemies had done a great deal more than that. They had started the second nuclear age. "These were nuclear weapons with a regional agenda, unveiled with a populist flourish," *New York Times* columnist Bill Keller wrote five years later. "And they had a religious subtext—the Hindu bomb, the Islamic bomb—that has become more acute as fundamentalists of the two religions gain ground in their respective countries."

India and Pakistan were the models for a new kind of nuclear power, Keller continued, and they were not alone. "North Korea is regarded as already nuclear. Iran is believed to be moving rapidly toward acquiring nukes. Libya and

Syria are watched with suspicion." (Libya has since abandoned its program.) And nuclear weapons in the hands of Iran, which prides itself on being Persian, not Arab, would likely inspire nuclear weapons ambitions in Egypt, Turkey, and even Saudi Arabia. Similarly, North Korea's nukes could eventually provoke responses in kind from South Korea, Taiwan, and even Japan, all of which understand the technology and are thought to have acquired at least some of it in the last three decades. Israel, surrounded by Iran and hostile Arab nations, and with its back to the Mediterranean, has stockpiled nuclear weapons for decades and has no intention of allowing them into enemy arsenals.

Iran is moving rapidly, indeed, and with expert help: impoverished physicists from ex-Soviet republics. Eduard Shevardnadze, then president of Georgia, said at a news conference in February 2003 that several nuclear physicists from the Sukhumi Institute of Physics and Technology in Abkhazia, an isolated republic on the Black Sea that fought for independence from Georgia, were working in the Iranian nuclear weapons program. Worse, the missing scientists seemed to have taken the institute's entire stash of highly enriched weapons-grade uranium with them. And if that weren't enough cause for alarm, radioactive cesium chloride, a powder that could be used in a so-called dirty bomb, was stolen from another part of the institute and was reportedly recovered in October 2002. The situation at the institute is "very dangerous," a senior member of Georgia's environmental ministry said. "We don't know what will happen in Abkhazia tomorrow." And that the Iranians had built a uranium enrichment facility of their own at Natanz, coupled with weapons-grade plutonium produced by a power reactor at Bushehr that was sold to them by the Russians, was also causing deep apprehension. So was impeding U.N. International Atomic Energy Agency inspectors in 2003 and 2004 while Teheran insisted its nuclear program was strictly to produce energy.

"Each new country that gets nuclear weapons multiplies the potential for a war involving a nuclear state," the Times's Keller observed. The first nuclear age was characterized by a stalemate between two industrial superpowers, he added, noting that both sides understood that a full-blown nuclear war would be so calamitous as to be out of the question. "The second is about insecure nations, most of them led by autocrats, most of them relatively poor, residing in rough neighborhoods, unaligned with and resentful of Western power." The old expression for them was *loose cannons*. These cannons, however, are atomic. And ironically, globalization seems to be hastening proliferation, making it, in his words, "just another unsavory but probably uncontainable technology, like

Internet porn." The use of nuclear weapons, as the boldfaced headline on Keller's article starkly noted, has now made the transition from unthinkable to eminently *thinkable*. "Poor countries can even finance their nukes by exporting other military material, as North Korea has done." Pakistan's nine-hundred-mile-range Ghauri ballistic missile (named after the twelfth-century Afghan warrior who conquered part of India), for example, is basically a North Korean Nodong. Pyongyang swapped the missile technology for Karachi's nuclear weapons assistance. And Pakistan has peddled its own missiles to Iran, Syria, Yemen, and other countries. A North Korean incentive for producing nuclear weapons is that they, too, are attractive to customers who see them as enhancing national security and prestige. And they are ideally suited to megaterrorism.

Then there is the Russian problem. Most of the roughly sixty thousand nuclear weapons of all types that were turned out by the time the Cold War ended are still around. So are the uranium enrichment facilities and stocks of plutonium that went into the weapons' cores. Some twenty thousand tactical and strategic weapons are at storage sites in the former Soviet Union, along with enrichment facilities and six hundred or so tons of highly enriched uranium and plutonium. That's enough to make about forty thousand bombs.

Finished weapons are less a problem than are highly enriched uranium and plutonium because the bombs and warheads are relatively well guarded and their nuclear cores are hard to remove. The raw fissile material, on the other hand, is guarded by the poorly paid successors to the once vaunted Red Army and their civilian counterparts. In hard times, principle can easily give way to survival and be rationalized on the ground that a society that allows its citizens to live in near squalor deserves no loyalty. As a consequence, Russia's porous south and the string of former republics that ring it from Kazakhstan to Kyrgyzstan to Tajikistan to Turkmenistan to Georgia have become a nuclear Silk Road. On April 6, 2001, Kazakh border guards seized two satchel-sized, lead-coated containers holding radioactive material officials would not describe. An American expert said he thought the shipment, which could have been bomb-grade, was headed for North Korea. It could very well have been headed for Al Qaeda and other terrorist groups, too, since they have financial resources.

Whatever the satchels contained was minuscule compared to what has been moving elsewhere. In May 1999 several grams of highly enriched uranium powder were intercepted on the Bulgarian-Romanian border. Five months later, plutonium was seized at Kara-Balta, Kyrgyzstan. In all, 550 reported incidents of illicit trafficking in radioactive material were reported worldwide between 1993 and 2001, though most transfers did not involve the weapons-grade variety. But on at least sixteen occasions they did. And the term *nuclear weapon* is a broad one. In June 2003 Thai police, acting on information from U.S. intelligence, arrested a forty-four-year-old Thai who was trying to sell as much as sixty-six pounds of cesium-137 for $240,000. Cesium-137 is a by-product of nuclear power plants and could also be used to produce a dirty bomb that would be set off by ordinary explosives that spread radioactivity. Experts were startled by the amount of the material, since only an ounce could make an effective bomb. The killing capacity of dirty bombs has been exaggerated, but they are valuable weapons for terrorists because of their capacity to terrify large numbers of people and cause chaos.

The situation is deeply worrisome, the more so because of the rapid proliferation of the ubiquitous Islamic terrorist groups, or jihadists. Milt Bearden, a thirty-year CIA veteran who was a senior manager for clandestine operations and served in Afghanistan during the Cold War, has warned that the capture of Osama bin Laden and the weakening or even the destruction of Al Qaeda will not end the terrorist threat. "The Islamic extremists who gathered around Mr. Bin Laden during the 1980s and '90s returned to their home countries where, like a cancer, they metastasized into the self-sustaining and deadly organisms that have since brought destruction to the United States and its allies as far afield as East Africa, Bali, the Philippines, Morocco, occupied Iraq and, now, Spain," he wrote soon after the savage attacks on the Madrid subway in March 2004. For these groups, too, weapons of mass destruction are the preferred instruments for inflicting maximum carnage.

"We are in a serious situation now: things have deteriorated," Rolf Ekeus, chairman of the Stockholm International Peace Research Institute and a recognized authority on superweapon proliferation, warned in May 2001. Alexander Schmid of the United Nations Terrorism Prevention Branch in Vienna echoed Ekeus: "There's an undeniable trend in proliferation [by] states. If this trend spills over into nonstate actors, we're in real trouble." It is and we are.

Less is known about the old Soviet Union's chemical and biological weapons program than about its nuclear research and development. But what is known is equally worrisome. In December 1992, a biologist named Ken Alibek, who was second in charge of the bioweapons program, defected to the West. He then related an incredible story about a vast enterprise that employed sixty thousand people in more than fifty laboratories, all dealing explicitly with biological weapons such as anthrax, Ebola, smallpox, yellow fever, and plague. Most of the laboratories were closed when the USSR collapsed, but large quantities of their product undoubtedly remain in inventory and are likely even more attractive to terrorists than fissile material. Equally worrisome, roughly forty thousand tons of chemical warfare weapons such as blister agents and nerve gas are also in inventory. Shchuch'ye, a town near the southern border with Kazakhstan, has a stockpile of almost 2 million artillery shells and hundreds of missile warheads that are filled with nerve gas and other deadly chemicals.

Middle Eastern and Asian terrorist groups would be eager buyers of the stuff. North Korea, which is chronically poor, would no doubt be an equally eager seller. So would impoverished guards at nuclear, chemical, and biological laboratories and storage facilities throughout Russia. That prospect is one of many that haunts Martin Rees. "If such were to obtain a nuclear weapon, they would willingly detonate it in a city centre, killing tens of thousands along with themselves; and millions around the world would acclaim them as heroes," he has written. "The consequences could be even more catastrophic if a suicidal zealot were to become intentionally infected with smallpox and trigger an epidemic."

The most insidious thing about chemical and biological weapons is that they are far cheaper and easier to produce than nukes, and in the case of communicable biological weapons such as anthrax, plague, cholera, typhoid, Ebola, smallpox, and yellow fever, they spread themselves.

The bioterrorist's weapon of choice, it turns out, would not be anthrax. It would be smallpox. "No human being is known to have been inflected by smallpox for close to a quarter of a century," Harold Varmus, a former director of the National Institutes of Health, wrote in a review of *The Demon in the Freezer,* a book about the disease. "Yet in a world sensitized to the dangers of terrorism, talk of smallpox is on every front page and at every dinner table. Because infection is often lethal, because the virus is easily transmitted between people, and because the discontinuation of vaccination three decades ago has made us highly vulnerable, it is not difficult to imagine an attack—it could be as simple as the undetected arrival in the country of recently infected, suicidal terrorists—that is truly terrifying." Varmus and Rees are on the same page.

The apparent end of naturally occurring smallpox was the reason all but two batches of the virus, which would be used to make vaccines, were officially destroyed years ago. One of the remaining batches is at the Centers for Disease Control and Prevention in Atlanta. The other is at the Vector Laboratory in Russia. "Officially," because other batches created by "dark biologists" could be stocked in a number of countries that are known to sympathize with the Islamic terrorist agenda. Richard Preston, who wrote *The Demon in the Freezer*, claimed that because bioweapons are so cheap and effective—they copy themselves in the human body—biologists have become what atomic physicists were in the 1930s and '40s: they are harnessing nature to produce a new kind of bomb.

With that in mind, federal health officials were seriously considering testing smallpox vaccine on toddlers and preschoolers in the autumn of 2002. Meanwhile, the Department of Defense was planning to do the same to half a million military personnel it was going to send into war against Iraq, and state health officials were asked to vaccinate workers who might come in contact with the victims of a terrorist attack. There was also recurring discussion about inoculating everyone. With all that talk, and with pictures in newspapers, magazines, and on television of masked health workers waddling in baggy airtight suits like creatures from another planet, Americans had become distinctly edgy by the end of 2002. It was therefore understandable for some people to react in panic when it was widely reported that a man from New Mexico had landed in Beth Israel hospital in New York with plague early that November. He contracted it at home, though, where it occurs naturally.

By then, the subject of terrorism had become a constant fact of daily life, both in North America and across Europe from Madrid to Moscow. There seemed to be almost daily news reports of Arabs linked to Al Qaeda being arrested in Germany, Italy, and the Netherlands, of small explosives going off in France, seemingly interminable security searches at airports, and unending, color-coded terror alerts. The warnings, credited either to the FBI or to the amorphous "intelligence community," were always too vague to act upon. The apparent idea, at least in the White House, was to keep the possibility of an attack out in public. That would create a win-win situation, since if there was no attack, the administration could use the better-safe-than-sorry argument, and if there was one, it could claim credit for being alert enough to have issued the warning. The frequent warnings were soon sharply reduced as the public started to ignore them. For their part, the nation's intelligence services were stung by accusations that they had failed to provide warning of the attacks on New York and Washington, so they strained to use communication intercepts and other information to keep the White House as thoroughly briefed as possible. Then, in late March 2004, the situation flared again when a former White House terrorism adviser published a book that charged President George W. Bush with ignoring ominous intelligence "chatter" in 2001 about possible attacks that were given to him at intelligence briefings. Richard Clarke, who advised four presidents on terrorism, made the revelations in *Against All Enemies.* "I find it outrageous that the president is running for re-election on the grounds that he's done such great things about terrorism," Clarke told *60 Minutes* on March 21. "He ignored it. He ignored terrorism for months, when maybe we could have done something to stop 9/11," an angry Clarke told CBS. Condoleezza Rice, Bush's national security adviser, responded by deftly claiming that her boss's initial response to terrorism followed that of his predecessor, Bill Clinton. Clarke also claimed that Bush and his inner circle were determined to connect Iraq and Al Qaeda despite the intelligence community's repeated assertions that no significant connection existed.

The destruction of the World Trade Center taught Americans a truth they never had reason to learn during the nuclear stalemate with the old USSR. Total civil defense is a myth. There is no shelter from an enemy determined to annihilate an opponent. The events of that terrible day demonstrated the victims' vulnerability and their government's inability to guarantee their safety from murderous, ruthless fanatics who measure their success by the amount of destruction inflicted and by body counts. That realization, which was also grasped by Americans' government, led inexorably to understanding that they were now targeted for weapons of mass destruction, including the biological kind. The possibility turned into deadly reality soon after the attack on the World Trade Center, when anthrax spores were sent in the mail, infecting twenty-two people and killing five.

For Al Qaeda and the rest of the burgeoning jihadist movement, the war against their sworn enemies around the world extends even beyond committing mass murder and destroying structures. It is a clash, in Samuel Huntington's apt description, between civilizations. That means culture, including art, is considered a fully justified target. This was the reason the Taliban systematically destroyed art that depicted human and animal forms and that for other reasons did not conform to what they believed, and continue to believe, to be the tenets of their narrow and inflexible theology. In March 2001 they caused gasps around the world by making a public display of dynamiting two enormous Buddhas that had taken monks decades to carve into the side of a mountain in Afghanistan fifteen hundred years ago. "All we are breaking are stones," the Taliban leader, Mullah Omar, remarked as the magnificent statues were blown up. The larger one was reduced to chunks of rubble that could be restored. But the smaller one, in the words of a distressed international rescue worker, was mostly turned to powder.

As horrifying as the destruction of the statues was, it was only the most apparent spectacle in a larger, relentless effort to eliminate all objectionable art. A month earlier, senior Taliban officials accompanied by armed religious guards had forced their way into the storeroom of the all-but-gutted National Museum in Kabul and, with stunned and disbelieving curators looking on, smashed every statue and other artifact that bore a human or animal likeness. "From afternoon until evening they broke statues," Omar Khan Masudi, the museum's eventual director, recalled with horror more than a year after the rampage occurred. "A few days later, they came back, and they followed the same procedure. They came back many times."

The Kabul museum, like its counterpart in Baghdad that was looted and ravaged two years later, was the victim of repeated clashes and warfare. It had held priceless Neolithic female figurines, stone Hindu goddesses in togalike robes, intricately carved Greek, Chinese, and Indian ivories, gold coins from the time of Alexander the Great, troves of early Islamic art, and more. Much original art had been lost a decade earlier when the building was gutted in the war against the Soviet occupation force. What had not been destroyed in the once incomparable archaeological museum was looted by desperately poor locals. If anything positive could be drawn from the awful episode, it was the imperative to protect cultural artifacts and make copies to be kept in safety elsewhere.

The terrible violence, both from Earth and its inhabitants, has given the two old Cold War enemies and much of the rest of the world common cause. In October 2002, for example, Chechen guerrillas fighting for the independence of their breakaway republic took over a theater in Moscow and began executing some of more than six hundred hostages before Russian troops burst in under cover of a debilitating gas and ended the murderous siege. The randomness and unpredictability of that hostage crisis evoked memories of the attack on the World Trade Center for many Russians, who came to believe that their country and its old enemy now faced the same threat. The subsequent murder of more than three hundred innocent schoolchildren and teachers underscored that belief. Terrorism in its several manifestations has forced the United States and the Russian Federation to take a Darwinian approach to survival: adapt or die. That means creating radical, and in themselves potentially dangerous, counterstrategies.

International relations are changing profoundly. The societies that are being stalked by the murderous fanatics are adopting doctrines similar to those of their enemies. Most fundamentally, that means an age of transnational response has begun in which there will no longer be privileged sanctuaries for terrorists. President Vladimir V. Putin, a veteran of his country's state intelligence agency, the old KGB, was explicit about the response within three weeks after the hostage crisis at the theater. "Russia will respond with measures that are adequate to the threat to the Russian Federation," he warned, "striking on all the places where the terrorists themselves, the organizers of these crimes and their ideological and financial inspirers, are. I stress, wherever they may be located."

President Bush quickly and publicly announced his support of Putin's strategy. That was not surprising, since seven months earlier it had come to light that the Pentagon had prepared its own new doctrine under the specific direction of the Bush administration. The report, called the Nuclear Posture Review, was delivered to Congress on January 8, 2002, and leaked to a *Los Angeles Times* defense analyst nine months later. Much of the plan was, and remains, classified. What came out, however, was a strategy that called for reducing the overall number of nuclear weapons but of significantly widening the circumstances in which they would be used. It put at least seven countries—China, Russia, Iraq, North Korea, Iran, Syria, and Libya—on a potential nuclear hit list. The war planners who produced the fifty-six-page document foresaw the weapons being used in three types of situations: against targets able to withstand nonnuclear attack; in retaliation for an attack with chemical, biological, or nuclear weapons; or "in the event of surprising military developments" that were not specified in the unclassified sections. The possibilities of using nukes in an Arab-Israeli war, because of an attack by North Korea against South Korea, or in a confrontation between China and Taiwan were specifically mentioned.

But the heart of the Nuclear Posture Review was about terrorism, both state-sponsored and freelance. It explicitly called for tactical nuclear weapons to be used in Afghanistan-type counterterrorist operations to penetrate and destroy heavily fortified underground bunkers, including those used to store chemical, biological, and nuclear weapons of any type. And it stressed the need for improved intelligence collection and targeting—both the mission of reconnaissance planes and satellites, in addition to a growing legion of old-fashioned spies and informers—and the ability to use precision weapons over long distances in combination with covert operations. The last has to do with special operations teams that would also collect intelligence, find targets, and fight.

The plan was roundly attacked as a setback to arms control. Testing the new, smaller nuclear weapons would contravene the voluntary moratorium on such tests, its opponents charged. It would also violate the spirit of the Nuclear Non-Proliferation Treaty by encouraging nations that do not have nukes to get them rather than remain defenseless against a possible U.S. attack. Under a headline that read "America as Nuclear Rogue," a lead editorial in *The New York Times* accused the Pentagon of trying to lower the threshold for using nuclear weapons and in the process undermine arms control. American military planners' behavior has traditionally "been tempered by the belief, shared by most thoughtful Americans, that the weapons should be used only when the nation's most basic interest or national survival is at risk," the editorial asserted. It is not hard to imagine those who crafted the Nuclear Posture Review, as well as those in the White House who encouraged them to do so, responding that a mass death warrant issued against Americans everywhere and threats of inflicting horrendous damage in the United States and elsewhere are indeed a threat to national survival. The Republican-controlled House Armed Services Committee rejected the plan sixteen months later because it amounted to a decisive setback for an already weakened arms control regime and set a dangerous precedent for potential copycats. But it would not die. With many congressmen favoring an approach to nuclear weapons that reflected post–Cold War realities, the White House and the Department of Defense continued to think about shifting from reliance on long-range ballistic missiles as a deterrent to smaller nukes that could be used against terrorists and rogue nations. "It is a return to looking at the defense of the nation in the face of a changed threat," Fred S. Celec, the deputy assistant to the secretary of defense for nuclear matters, was quoted as saying. Opponents, including Senator Edward M. Kennedy and Senator Dianne Feinstein, charged that, given the Bush administration's insistence that other nations abandon nuclear weapons altogether, the push for smaller nuclear weapons was ill-conceived and reckless. Feinstein went so far as to call it "diabolical."

The proliferation of fissile material, and nuclear, chemical, and biological weapons themselves, particularly in the hands of terrorists, carries potentially disastrous consequences. But this is only one among many dire threats posed by high technology. Looking further ahead—though not too much further— some heavy thinkers have predicted a potentially more insidious danger, and unlike nuclear weapons, it is one that would be out of humanity's control. Richard Feynman, the Nobel Prize–winning physicist, predicted in 1980 that the extreme miniaturization of computers that mimic life, including repro-duction, loomed as a danger to humanity. More recently, Ray Kurzweil, a pro-lific and technologically literate futurist and entrepreneur, has looked far ahead and seen the ultratiny computers that mimic life as a force for both good and evil. He has predicted that nanotechnology (from the Greek *nanos*, or dwarf) will create self-replicating machines on the atomic level that will have the capacity to improve life. They will do that by turning themselves into solar cells that replace fossil fuels, for instance, or by entering the bloodstream to supplement natural immune systems and fight cancer, pathogens, arterial plaque, and other threats to life. If cancer is an army of killer cells that grows wildly by attacking and feeding off other cells, then the self-replicating "nanobots" (nano-robots) can be programmed to hunt and destroy cancer cells the way military special forces hunt and destroy the Taliban and other mur-derous fanatics. They could also be used to reconstruct body parts and much more.

They will indeed become amorphous armies, and like other armies they will have the capacity for both good and evil. The microscopic creatures will have potentially dangerous capabilities. If nanobots cannot replicate them-selves, Kurzweil explained in a book called *The Age of Spiritual Machines*, nanotechnology will be neither practical nor economically feasible. Auto-matic breeding is therefore essential. But, he warned, a software failing or other problem that cannot stop them from replicating at a certain point, ei-ther by accident or by design, would unleash a nanobot population explosion that could "eat up everything in sight." In other words, an exponentially ex-ploding nanobot population would devour organic matter rich in carbon. That includes us. We would be in their food chain, along with other mam-mals like cattle, pigs, and sheep. "They could eat up everything in sight," Kurzweil has written. It would be the ultimate locust plague, though unlike the insects, the nanolocusts would feast on matter rich in carbon. While an accidental plague of nightmarish proportions is remotely possible, he contin-ued, the odds against its occurring are high. Kurzweil is no alarmist. He went on to note that hundreds of nuclear reactors have operated for decades and there have been only two dangerous accidents, both in the Chelyabinsk re-gion of the former Soviet Union. Kurzweil also noted that there are tens of thousands of nuclear weapons stashed on the planet and not one has ever ex-ploded accidentally.

That having been said, Kurzweil turned to the greater danger: the deliberate, hostile use of nanotechnology for war or terrorism. And he was not necessarily referring to using the technology in an act of self-destruction. "It is not the case that someone would have to be suicidal to use such weapons. The nanoweapons could easily be programmed to replicate only against an enemy; for example, only in a particular geographical area. Nuclear weapons, for all their destructive potential, are at least relatively local in their effects," he warned. "The self-replicating nature of nanotechnology makes it a far greater danger." The idea of swarms of nanoparticles threatening their creators so appealed to Michael Crichton, the prolific author of techno-thrillers, that he made it the theme of a novel called *Prey*.

K. Eric Drexler, who first sounded the alarm about nanotechnology in 1986, has become more sanguine. That year he described a powerful micro-manufacturing system, which he called the assembler, that would use robots the size of bacteria to combine individual molecules into products. But he warned that that the robotic molecular-manufacturing system could be programmed to copy itself and spread out of control. This would produce what was popularly called the dreaded gray goo. But eighteen years later, Drexler coauthored a paper in the journal *Nanotechnology* explaining that nanotechnology for manufacturing does not have to rely on dangerous self-replicating machines; that nanotech-based fabrication can in fact be thoroughly nonbiological. No biology, no out-of-control swarms of microscopic killer bugs (at least by accident).

Bill Joy, like his friend Ray Kurzweil, worries that humanity could become so dependent on the machines that it would become subservient to them. Joy cofounded Sun Microsystems and is another computer expert. He has warned that humans could over a long period gradually drift into such dependence on thinking machines that a time would come when they have no choice but to accept the machines' decisions. "Eventually, a stage may be reached at which the decisions necessary to keep the system running will be so complex that human beings will be incapable of making them intelligently. At that stage, the machines will be in effective control. People won't be able to just turn the machines off," he added grimly, "because they will be so dependent on them that turning them off would amount to suicide." And even if humans continue to control machines indefinitely, it will be a tiny, elite group that will therefore have enormous control over the masses. And the masses themselves will have become superfluous, a useless burden on the system, and therefore perhaps to be disposed of by a ruthless elite.

Martin Rees, taking a wider view of the dangers facing humanity, sees potential catastrophe everywhere. And he has traced his deeply troubled worldview at least as far back as H. G. Wells, who gave a respectful lecture on science at the Royal Institution in London in 1902. Wells predicted that the twentieth century would see more innovation, more scientific advance, than had been accomplished in the previous millenium. But the author of *The Shape of Things to Come, The Mind at the End of Its Tether,* and other novels that increasingly reflected his ambiguous, troubled feelings about science, looked into the future and became pessimistic. "It is impossible to show why certain things should not utterly destroy and end the human race and story; why night should not presently come down and make all our dreams and efforts vain," Rees has quoted him as saying. Wells envisioned dire threats everywhere: "something from space, or pestilence, or some great disease of the atmosphere, some trailing cometary poison, some great emanation of vapour from the interior of the Earth."

In *Our Final Hour,* Rees took it from there. "Humans already have the wherewithal to destroy their civilization by nuclear war; in the new century, they are acquiring biological expertise that could be equally lethal; our integrated society will become more vulnerable to cyber-risks; and human pressure on the environment is building up dangerously," the cosmologist warned. "The tensions between benign and damaging spin-offs from new discoveries, and the threats posed by the Promethean power science gives us, are disquietingly real, and sharpening up."

Sharpening up, for Rees, means that the lines representing the availability of weapons of mass destruction and the lessening number of people required to use them are crossing. That is what he meant when he referred to the crazed carrier of smallpox who knowingly starts an epidemic. Even terrorist networks would be overstaffed in the nightmarish world of what Rees calls megaterrorism. The ultimate dirty deed could be accomplished by "just a fanatic or social misfit with the mindset of those who now design computer viruses. There are people with such propensities in every country—very few, to be sure, but bio- and cyber-technologies will become so powerful that even one could well be too many."

To Kurzweil and Rees, add Stephen W. Hawking, the brilliant theoretical physicist whose popular book on astrophysics, *A Brief History of Time,* was long a bestseller. For Hawking, too, the attacks on the World Trade Center and the Pentagon marked an ominous change in the fate of Earth. He worried in particular about the evil dimension created by biology's dark side in the service of terrorists and was moved to warn a month after the attacks that civilization would not survive the next thousand years unless it spread to space. "There are too many accidents that can befall life on a single planet," he told a British journalist, adding that Armageddon threatens not from a Cold War–type nuclear holocaust so much as from silent, insidious microbes. "In the long term, I am more worried about biology. Nuclear weapons need large facilities, but genetic engineering can be done in a small lab. You can't regulate every lab in the world," he warned. In fact, even facilities such as centrifuges for making bomb-grade fissile material have now been miniaturized. Nor are superweapons all that Hawking worries about. He told Britain's Press Association in September 2000 that Earth's atmosphere "might get hotter and hotter until it will be like Venus, with boiling sulfuric acid. I am worried about the greenhouse effect." Humans might not survive this millennium, he said, unless they migrate to other places. "It takes too many resources to send each person into space," Hawking added, "but unless the human race spreads into space, I doubt it will survive the next thousand years."

Kurzweil, Rees, and Hawking are world-class scientists who have tried to project the hard reality of their disciplines into the future; to describe the probabilities of the dangers they believe are lurking ahead. They would abhor being called futurists. Yet in a sense, that is what all who try to describe what they believe lies ahead are, even when their apprehension is grounded in hard science and objective probability. The French poet and critic Paul Valéry noted with eloquent simplicity, "We hope vaguely, but dread precisely."

And precise prediction is notoriously wrong. History is littered with broken crystal balls. It was taken almost as a matter of faith by many who considered themselves knowledgeable about international conflict, for example, that at least one, and possibly as many as five, nuclear weapons would be used between the end of World War II and the beginning of the new century. Although President Harry S. Truman told reporters at a news conference "every weapon that we have" was a candidate for use in the Korean War, and that obviously included the atomic bomb, he refused to allow one to be dropped. And no pundit would have predicted that the Union of South Africa would voluntarily have dismantled its six nuclear bombs, even after a peace of sorts had been declared with neighboring Angola. But it did. And both Argentina and Brazil, bitter political rivals in the 1960s and 1970s, briefly flirted with obtaining nuclear weapons, then backed off. The dreadful risk, they decided, was not worth an illusionary benefit. That was counterintuitive for the weapons prognosticators.

So was the failure to fight World War III. David Baker, a British space expert who held a doctorate in earth and planetary sciences and a diploma in astronautics, wrote a book called *The Shape of Wars to Come* that appeared in 1981, in the throes of the Cold War. In it, Baker claimed that the Third World War would start in space and could involve particle weapons of such immense power they would strike Earth's surface with a radioactive cone in which everyone would die: "Every living thing would perish and mutants would emerge from lesser orders of life capable of surviving the irradiation. The use of such a terrifying weapon would be the ultimate application of enhanced radiation devices." Baker was not saying the use of such a terrible device was a certainty, but he made it a prime candidate for application during the next major war; an all-out war in space and on Earth. But the fantastic (and fantastically expensive) particle weapons were never built, let alone orbited, and the space war they were allegedly going to fight never happened. Baker's publisher was less restrained than he was. "The next World War will start in space," the jacket's flap copy warned. "How long before it starts? If David Baker is correct in his predictions—and few people are better qualified to write on this subject—this book could be one of the last things you read." Whew.

Nostradamus is the patron saint of the Third World War. Or at least he was while the two superpowers stood "on the brink," as they said in those days. The sixteenth-century astrologer and physician was credited by one of his interpreters as predicting that nuclear weapons would be used against the Soviet Union; that Russian and other Warsaw Pact armor and infantry would roll into Austria, West Germany, Switzerland, and would even occupy Paris; that France itself would suffer an airborne invasion in 1999; and that ballistic missiles would obliterate Japan and the West, starting a Third World War that the "Reds" would win. The soothsaying appeared in a Nostradamus book that came out in 1983, when nuclear war was plausible, at least technically.

By 1998, with the former Soviet Union and several of its old allies picking themselves up and brushing themselves off after the fall, references to a climactic world war fought with nuclear weapons, as predicted by the famous forecaster, quietly evaporated. (And Hal Lindsey's dire prognostications were radically modified to conform with reality.) In fact, *The Mammoth Book of Nostradamus and Other Prophets* even credited the old boy with foretelling the demise of Soviet Communism (page 424). And like others who claimed to divine what Nostradamus had allegedly predicted, Damon Wilson mostly stuck to safer ground: Pearl Harbor, John Kennedy's assassination, World War II, the atomic bomb, the appearance of Fidel Castro, the Hungarian revolution, and even the war between Iran and Iraq. The author took "From the sky will come the giant king of terror" to mean that doomsday would dawn in July 1999. Yet another student of Nostradamus, whose own tome was published in 2001, also stayed away from World War III, preferring instead to lay heavy emphasis on Napoléon and to note that an "arrow" (really a comet) was due to slam into the Aegean Sea on or about August 19, 2004.

To his lasting credit, Nostradamus also seems to have predicted there would be a space program, at least in the opinion of those who have pored over his work and believed it to be prescient. If he thought it is humankind's destiny to have a space program, which has of course come true, then he would undoubtedly have envisioned people going to space and living there.

Martin Rees and Stephen Hawking, among others who have thought about Earth's tenuous existence, believe that colonizing off the planet is imperative. Rees has grimly described the multiple dangers technology will bring to this civilization. But he has also gone on record as explaining that technology can also deliver the means of salvation.

Although Rees has calculated that there is trouble ahead, he makes no pretext that it is certain, nor that it will end in catastrophe. But he has expressed a strong belief that it is nevertheless wise to expand humanity's base: "Even a few pioneering groups, living independently of Earth, would offer a safeguard against the worst possible disaster—the foreclosure of intelligent life's future through the extinction of all humankind." Hawking agrees. In spite of his own conviction that grave danger inevitably lies ahead, he has called himself an optimist who firmly believes "we will reach out to the stars."

But also to Earth, with the stars as a kind of backup. The final, perhaps ultimate, irony is that the human race is far better at controlling, or at least ameliorating, the deadly and destructive forces of nature than it is at controlling its own dangerous science, technology, and deadly impulses. Enlightened science has provided the means to survive the wrath of the cosmos, which flings asteroids and comets at Earth and tries to bathe it with poisonous radiation from solar flares and places beyond the Sun. The planet is an intensely living thing that inflicts upon itself earthquakes and volcanic eruptions, fearsome weather, deadly epidemics, and other assaults. The death and destruction humans bring upon themselves, as well as on the only world that can nurture and protect them, is quite another, more pernicious, problem.

There is, however, a way to significantly reduce the threat from both nature and civilization's most disagreeable and dangerous children. It is to use space to enhance Earth's survival and at the same time spread the seed elsewhere. A "program" is generally thought of as being a plan or system under which action may be taken toward a goal. By that definition, there is no space program, since the space agency has no overarching, articulated goal. But there is a worthy one: protecting Earth and life on it. That means focusing our space resources not only on fending off dangerous intruders from elsewhere and reducing the effects of subterranean violence, but on increasing the food supply; producing clean energy, containing and then reversing pollution and global warming; preventing drought and minimizing weather damage through civil engineering and better forecasting; fighting naturally occurring diseases; planning communities that are compatible with the environment; monitoring and protecting resources (overfishing and deforestation being only two obvious dangers); backing up civilization's cultural, scientific, political, and other records for safekeeping at protected sites on Earth and in a large lunar colony; creating populated stations in space; and reducing or stopping the spread of weapons of mass destruction, the systems that deliver them, and the terrorists who want to use them to commit more atrocities. A great deal of this massive remedial undertaking would certainly have to be done on Earth, not in space. But the most important fact of this century is not that Earth is threatened in many ways. It is that for the first time in all of its history a decisive means of protecting the home planet exists. It is by using space. And it is already well under way.

# THE ONCE AND FUTURE
# SPACE PROGRAM

A specter is haunting NASA. It is that of an agency without a mission. But there is a mission: using space to protect Earth.

Once there was a plan for reaching beyond the atmosphere. And there was a dream. They were conceived for either of two basic reasons: adventure (including exploration and scientific discovery) or safety. Understandably, the adventurers led the way, first in fiction and then in fact. Lucian of Samosata, a second-century Greek satirist, became one of the first known space-fiction writers when he published *Vera Historia*, or *True History*, part of which described his hero flying to the Moon by taking off from Mount Olympus using one wing from an eagle and another from a vulture. Icaromenippus, the scientifically curious adventurer (whose relation to the Icarus of Greek myth was not coincidental), ultimately stood on the lunar surface and took in planet Earth, with sunlight glittering off its vast ocean.

Jules Verne's *From the Earth to the Moon*, which was printed in 1865, closely followed by *Round the Moon*, told the story of a small party of adventurers who were shot to the Moon by a giant cannon as a scientific experiment sponsored by the Gun Club of Baltimore. Like many science fiction writers who followed him, Verne was short on character development and long on technology, which he studied voraciously. He understood, for example, that the spinning of the Earth west to east is fastest at the equator and therefore increases the velocity of whatever is launched from there. So he put his cannon, the Columbiad, at Tampa Town in Florida. That, it was to turn out, is directly across the state from what was to become Cape Canaveral and the Kennedy Space Center. And like astronauts and cosmonauts, they were flung toward the east to take advantage of Earth-spin. The vivid foray to the Moon by the intrepid thrill-seekers fired the imaginations of three of the space age's towering pioneers: Konstantin E. Tsiolkovsky, a Russian, Hermann Oberth, a German, and Robert H. Goddard, an American.

In 1869, four years after *From the Earth to the Moon* appeared, the *Atlantic Monthly* ran a three-part series called "The Brick Moon," which carried no byline but was written by Edward Everett Hale. The respected author rejected Verne's monster cannon for propulsion, probably because he knew that the sudden, tremendous acceleration would turn his moon's 12 million attached bricks into pebbles and dust (as it would have turned Verne's adventurers, Ardan, Barbicane, and Nicholl, into what the British would call strawberry jam). Instead, Hale had the project's designers use two huge flywheels that would accelerate the hollow moon gradually and finally fling it into orbit. One of the key participants, who chronicled the story, was a tycoon named Captain Frederic Ingham. He spoke for many readers and their spiritual descendants when he wondered, "For myself, to this hour, I never enter board meeting, committee meeting, or synod, without the queer question, What would happen should any one discover that this bearded man was only a big boy disguised?"

That appellation applied to the true father of rocketry and astronautics as well. Konstantin Eduardovich Tsiolkovsky was born in Izhevskoye, a small town southwest of Moscow, in 1857. A happy childhood turned painful when an attack of scarlet fever severely diminished his hearing, forcing him to use a tin ear horn, which was deeply humiliating. He would write years later that his deafness made every minute spent with other people "torture." Then, at thirteen, his beloved mother died. His father was an itinerant forester and clerk who had little money and no time to help his son with schoolwork or much else. Tsiolkovsky therefore turned inward. He not only pored over the few mathematics, physics, and astronomy books he could find in the backwater where he lived, but constructed his own experiments, including steam engines, windmills, and pumps. Astronomy—wandering around other worlds—got Tsiolkovsky off Earth to where there were no classmates making fun of the large tin funnel he held to his ear. After a couple of teaching stints in the region, the impoverished sixteen-year-old went to Moscow and virtually moved into the public library, which was of course stocked with science and mathematics books. And there he came under the influence of a Christian theorist and mystic named Nikolai Fyodorov, who had many radical theories. He believed that Earth is not humanity's natural home, that everything in the universe is alive, that everyone who ever lived is out there somewhere (including

Tsiolkovsky's lost mother), and that as creation's most intelligent creatures, humans have a responsibility to bring order to the chaos of the cosmos.

Fyodorov got his young disciple thinking so seriously about flying through the cosmos that he turned him into a self-described gravity-hater, since gravity imprisoned him on Earth. Tsiolkovsky therefore invented a machine that would break gravity's tenacious grip: the rocket. And the prodigy's rocket was not the magical and mysterious device of fantasy, but one based on the reality of chemistry, physics, and mathematics. He calculated that a rocket would have to reach a velocity of roughly eighteen thousand miles an hour to break Earth's gravitational hold, and that two or more rocket stages, each falling off when its fuel was expended, would be necessary to reach an altitude where it would circle Earth. Tsiolkovsky even theorized that hydrogen would produce by far the most energy relative to its light weight. All of this and more was described in his seminal work on rocketry, "Exploring Space with Reactive Devices," which was published by the journal *Scientific Review* in 1903. The Wright brothers took to the air at Kitty Hawk the same year.

The great Russian visionary's imagination, perhaps reflecting Fyodorov's teaching, also extended to works of fiction. They, too, set the stage for the space age. In *Beyond the Planet Earth*, written in 1896 and finally published in 1920, he had six men—a Frenchman, an Englishman, a German, an American, an Italian, and a Russian—become disillusioned with existence on Earth and the decadent and boring "pleasures of life." So they applied their love of science to colonizing space. Tsiolkovsky was no Verne. Although it was ostensibly a science fiction novel, *Beyond the Planet Earth* was really a thinly disguised physics treatise on moving to and living in space. It described in some detail an Earth-circling space station, a space suit, landing on and mining the Moon and an asteroid, colonies orbiting Earth at a distance of 20,500 miles so each orbit took twenty-four hours, and the closed biospheres they would require to sustain life.

"We are now in orbit around the Earth in this rocket ship, at a distance of 1,000 kilometers, and making a full circuit every 100 minutes," the spacefarers reported on April 10, 2017. "We have constructed a large conservatory in which we have planted fruit and vegetables. These have already given us a number of harvests, thanks to which we are well fed, lively and healthy, and our needs are fully ensured for an indefinite period." It was the first real blueprint for spreading the human seed off Earth.

Tsiolkovsky in essence drew the first blueprint for going to space, and those who followed filled it in with ever-greater detail. Oberth, who had studied medicine at the insistence of his father, did pioneering work on how the human body would react to the stresses of rocket propulsion and a gravity-free environment. Goddard, a reclusive inventor, launched the first liquid-propelled rocket in Auburn, Massachusetts, on March 16, 1926. Meanwhile, pulp fiction was glorifying space travel (and the creepy creatures that lived there), and rocket societies were springing up in the United States, Great Britain, France, the Soviet Union, and elsewhere. Chief among them was the Verein für Raumschiffahrt, or the Society for Space Travel, which started in Germany in July 1927. The VfR, as it was called, attracted a small galaxy of talented engineers and scientists, most notably Oberth. And in 1933, an eighteen-year-old aristocrat and engineering prodigy joined the club. His name was Wernher von Braun.

The Treaty of Versailles prohibited Germany from producing such conventional weapons as submarines, warplanes, artillery, and tanks. It did not mention rockets, however, because its framers' knowledge of them had to do with celebrating the Fourth of July, Bastille Day, and other holidays. No thought was given to their use as weapons. But the German army knew better. It saw the rockets that were designed and fired by the VfR as potentially very long-range artillery. The generals therefore bankrolled the always hard-up rocketeers, who obligingly turned model rockets into increasingly sophisticated missiles. Wernher von Braun and a rocket team were eventually given an extensive rocket-development facility at a place called Peenemunde, on an island in the North Sea, and there they developed the A-4 (later V-2), a forty-six-foot-long ballistic missile that could carry a 2,200-pound explosive two hundred miles. The A-4's first successful test was on October 3, 1942. Watching it lift off for the first time, Krafft Ehricke, one of its designers, called it a "fiery sword in the sky." And so it was. V-2s were used against Belgium, France, and the south of England during the latter stages of the Second World War. They had no effect on its outcome. But V-2s were important because they were the first long-range ballistic missiles and, as such, were the true ancestors of the nuclear-armed ballistic missiles that hid in underground silos and submarines during the Cold War and afterward, as well as the rockets that carried men and women to space and robot explorers to distant worlds.

The German rocket team, foremost among them von Braun, surrendered to the U.S. army as the war ended and were rushed to the United States in a project code-named Paperclip. Before von Braun left Europe, he was asked to write a report on what he saw as the future of rocketry. He soon handed over his "Survey of Development of Liquid Rockets in Germany and Their Future

Prospects," which portrayed the A-4—"known to the public as V-2," von Braun cannily noted—as the true beginning of the realization of space travel. "We are convinced that a complete mastery of the art of rockets will change conditions in the world in much the same way as did the mastery of aeronautics and that this change will apply both to the civilian and military aspects of their use," the master rocket designer and visionary predicted.

The U.S. army, like the vanquished *Wehrmacht* and its new opponents in the Union of Soviet Socialist Republics, could not have cared less about exploring Mars. It saw the V-2 as a weapon with immense potential to destroy enemy targets far behind the lines, and it therefore put von Braun and the others to work creating advanced ballistic missiles. They were developed in the late 1940s and early 1950s and were called Redstones and Jupiters.

Whatever the respective military establishments on either side of the Iron Curtain wanted, however, the dream of heading for space was enhanced by the development and production of ever more sophisticated rockets, even those invented for war. And von Braun and several of his associates, with others, readily answered the call of space as well. Exploring space was exciting and interesting, after all. Designing devices that exploded was neither.

Back in 1944, as V-2s launched from occupied France pummeled the south of England, G. Edward Pendray, a founding member of the American Rocket Society, and Jacques Martial published an article in *Collier's* magazine predicting "superstratosphere" airplanes that would take off using turbojets and have rockets kick in at an altitude of ten or fifteen miles. They foresaw passengers making the trip from New York to Paris in three hours. "It will be quite possible," they predicted, "for an American businessman, in a hurry, to have breakfast at home, lunch and a conference in London, and be back home again for dinner." That dream was realized, and ultimately abandoned as uneconomical, with the Anglo-French Concorde supersonic airliner.

A year later, Waldemar Kaempffert, a former editor of *Scientific American* and then the science and engineering editor of *The New York Times*, envisioned rockets saving humanity in a book called *Science Today and Tomorrow*. Astrophysicists, he wrote, were predicting that Earth would eventually turn into a "cold cinder swimming around the sun," its atmosphere gone, its oceans and lakes dried up. "Must the last man die of starvation and thirst? Possibly the rocket ship is man's last hope. By the time the earth has become senile and unlivable, Venus will be ripe for intelligent beings. So it may happen, aeons hence, that Venus may be colonized by the earth as America was once colonized by Europe. And the earth will wheel around its orbit, an abandoned, planetary wreck of its former luxuriant green self." Nobody would voluntarily land on a planet whose atmosphere is hot enough to melt lead, but Kaempffert had no way of knowing that when he wrote his book. And comparing the colonization of Venus to that of America was a grossly misleading analogy. But he very likely got the rest right.

The future's defining moment came, or at least seemed to come, in October 1951 when the First Annual Symposium on Space Travel was held at New York's Hayden Planetarium. The meeting brought together six experts with diverse knowledge whose contributed papers would collectively form the basis for a cohesive, evolutionary, very long-range, national space program. Expanded versions of the presentations appeared in eight installments in Collier's between 1952 and 1954 under the general title "Man Will Conquer Space Soon." The series, as well as a book called Across the Space Frontier, which combined the articles, was illustrated by two technical illustrators and Chesley Bonestell, a talented space artist whose vivid planetary scenes had a real-world basis instead of being purely imaginary.

In the articles and the book, as at the meeting in the planetarium, Joseph Kaplan, a professor of physics at UCLA, described the space environment as it was then known. Heinz Haber, also from UCLA and a specialist in space medicine, explained the physiological challenges of spaceflight. A United Nations lawyer named Oscar Schachter made the point that no nation had a claim on space and that, like the high seas, it belonged to "all mankind." Fred L. Whipple, chairman of the Department of Astronomy at Harvard, described the big picture: the Moon, Mars, Venus, and other planets as destinations for spaceships from Earth. And the means to accomplish that—the hardware for the exploration of space—was laid out by von Braun and Willy Ley, another VfR alumnus and author, most notably, of a hefty space bible called Rockets, Missiles, and Space Travel, which was itself a pioneering work on the whole subject of astronautics. Ley's paper described a 250-foot-in-diameter space station that would rotate to create artificial gravity and circle Earth every two hours at an altitude of 1,075 miles.

The heart of the plan, as described by von Braun, constituted the essential elements of what generally is considered to be a modern, detailed refinement of Tsiolkovsky's blueprint for humanity's move to space. It started with the building of a sleek, winged shuttle that would be launched by a huge two-stage, reusable rocket. The shuttle's purpose was to carry prefabricated parts of the station to orbit for assembly. The completed station would in turn circle Earth in a near-polar orbit, meaning most of the planet that rotated under it would be subject to continuous observation. The shuttle would therefore also be a manned observation satellite. The station was also to be the embarkation point for expeditions to the Moon and then to Mars.

No one in the fraternity doubted that the enigmatic Red Planet would be visited and then colonized by earthlings. The year that important meeting took place in the planetarium, 1951, a largely unknown writer named Arthur Charles Clarke, who had an honors degree in physics and chemistry from King's College, London, and who was chairman of the British Interplanetary Society, published a book called *The Exploration of Space*. In it, the future co-writer (with Stanley Kubrick) of the movie *2001: A Space Odyssey* and the author of a small mountain of space books, factual and fictitious, predicted that Mars would and should be studded with several "pressure domes" or bubbles connected by air locks in which people would live exactly as they do on Earth. The dome idea anticipated the self-contained Biosphere 2 experiment that was started in Oracle, Arizona, north of Tucson, in September 1991. The developers of that problem-plagued operation—oxygen had to be pumped in seventeen months after eight people entered it—called it Biosphere 2 because they considered Earth Biosphere 1. The idea was to create a closed, self-sustaining system that would simulate Earth's ecology.

One closed dome in *The Exploration of Space* was illustrated in color. The huge, transparent bubble contained ordinary buildings of various shapes and sizes, automobiles, and people going about their business in a city that might have been Minneapolis or Rochester. The area immediately surrounding the dome was verdant and gave way to a reddish desert and jagged mountains. Clarke later repeated the theme in a novel, *The Sands of Mars*, which came out in 1967, and in *The Promise of Space*, a nonfiction work that was published the following year. So yet another writer with impeccable scientific credentials projected human civilization on Mars, and possibly on Venus as well (Clarke, too, could not have known about the intense heat under the thick Venusian atmosphere).

The plan to move people from Earth to space was incremental, logical, and progressive, like the development of aviation. Cornelius Ryan, a *Collier's* editor who attended the symposium and who later wrote *The Longest Day*, the epic story of the Normandy invasion in 1944, wrote the introduction to *Across the Space Frontier*. He made the point, right at the beginning, that successively faster aircraft were leading the way to space. He mentioned test pilot Bill Bridgeman's record-breaking flight to 79,494 feet—the "border" of space—in a Navy Skyrocket, also in 1951. That flight had, in turn, been made possible four years earlier when Major Chuck Yeager became the first to fly faster than the speed of sound in an X-1. Ryan also noted that a secret successor to the Skyrocket, capable of ascending to thirty-seven miles at eighteen hundred miles an hour, was being built. Whatever it was, it was ultimately surpassed by the rocket-propelled North American X-15, which set an altitude record of sixty-seven miles and a speed record of 4,520 miles an hour. "These efforts being put forward today in the development of research planes and high-altitude rockets are all directed toward one goal: the manned rocket craft of the future." The altitude record was broken forty-one years later by a manned rocket called SpaceShipOne. "Man is reaching out beyond the atmosphere and, in the words of Wernher von Braun, the conquest of space is 'as inevitable as the rising of the sun.'"

What was also taken to be inevitable by Ryan and other patriotic space enthusiasts was relentless competition with the Soviets for the control of space. He warned that the Kremlin had an extensive missile and rocket program that used tremendously powerful launch vehicles. "One of their top scientists, M. K. Tikhonravov, a member of the Red Army's Military Academy of Artillery, let it be known that on the basis of Soviet scientific development such rocket ships could be built and that the creation of a space station was not only feasible but definitely probable." And *Destination Moon,* a film that also came out in that charged year of 1951, and which was based on a Robert A. Heinlein novel, sounded the same warning. In it, a retired general named Thayer, who was trying to talk a roomful of corporate executives into financing a private moon rocket, warned about the abidingly sinister Red threat, "We are not the only ones who know the Moon can be reached. We're not the only ones who are planning to go there. The race is on and we'd better win it because there is absolutely no way to stop an attack from outer space. The first country that can use the Moon for the launching of missiles will control the Earth. That, gentlemen, is the most important military fact of this century." It was compelling drama. But it was absolutely wrong for at least two reasons. Building a missile base on the lunar surface would have been so extraordinarily expensive that neither side could have afforded it. What is more, it would have taken the missiles three days to reach the United States, under observation all the way, and therefore giving the Pentagon far more time to retaliate than if the missiles had made the thirty-minute trip over the Arctic from Siberia.

But Soviet ballistic missiles were real enough. The Red Army had captured a number of V-2s at the end of the war, along with one important scientist and hundreds of engineers and technicians. From this "starter set,"

Sergei P. Korolyev, soon to be known in the West as the mysterious and enigmatic Chief Designer (his identity was carefully protected to prevent a possible CIA assassination), Mikhail Tikhonravov, and other talented and imaginative engineers created their own rockets. Like their American counterparts, and contrary to what their political and military benefactors demanded, they were also obsessed by an urge to send their race to space and to explore other worlds. They knew that for the first time in history, they could create machines to realize those ancient, transcendental goals. The drive to do so was overwhelming. But also like von Braun and their other competitors, they were under explicit orders by the government that funded their research to turn their talents to the development of weapons to protect the motherland, not to irrelevant and wasteful machines that took pictures of Mars and Venus. Korolyev was patiently humored by his benefactors in the Kremlin until reaching those two planets began to figure in Cold War propaganda. Then some money was found, but never enough to do the missions correctly. Vasily P. Mishan, who succeeded Korolyev, later said in frustration that they were competing against the Americans with "wooden rubles."

The competing designers handled the dilemma the same way out of necessity. They turned out ever larger and longer-range ballistic missiles that could carry weapons with terrible destructive power over great distances. Meanwhile, they quietly but steadfastly pursued the dream at which their political and military patrons sneered. Like the multifaceted but highly integrated program in the United States, as shown at the Hayden Planetarium meeting, a number of conferences were sponsored by the USSR Academy of Sciences that were devoted to a single aspect of spaceflight: cosmic radiation, for example, or the magnetic field, the ionosphere, and so on. Tikhonravov later wrote that by 1953 Soviet rocketry had advanced to the point where it was possible to plan an Earth-satellite program. The satellites were named Sputnik, or "fellow traveler," and the huge launchers to carry them were specially adapted ballistic missiles called R-7s. In November 1956, Korolyev, Tikhonravov and their colleagues began designing a Sputnik that could carry a man and calculating what it would take to land a spacecraft on the Moon. At the same time, Korolyev was pushing his designers to create a simple but workable satellite that could be launched as part of the Soviet Union's contribution to the International Geophysical Year in 1957 and 1958. The Russians duly announced that fact. But no one took it seriously. They claimed to have flown the first airplane near St. Petersburg in the early 1880s, after all, and even to have invented baseball. So they were taken in the United States and elsewhere in the West to be blustering buffoons who had a well-founded inferiority complex.

That changed on October 4, 1957, when an R-7 thundered off the concrete launch pad at Tyuratam, on the Kazakh steppe, and tossed a 186-pound Sputnik into orbit. The news was broken in a routine Radio Moscow announcement three hours after the shiny sphere's launch. It flew over the United States twice, racing smoothly past its starry background, before anyone realized it was there. But when they did realize, a storm of self-flagellation and recrimination broke, much of it directed at President Eisenhower, who was widely portrayed as an amiable but distracted man who was more interested in golf than national security. That was anything but the case, as his important role in aerial and space reconnaissance showed. Ike had authorized U-2 flights over the USSR the previous year, and the first space reconnaissance satellite was developed on his watch. He gamely answered his critics by insisting—correctly—that the United States did not trail the Communists in science and technology. The president, the National Security Council, and the Joint Chiefs of Staff were in fact deeply troubled. But it wasn't about the Sputnik. They were concerned about the R-7. A rocket that could fling a 186-pound machine into orbit at a velocity of more than seventeen thousand miles an hour would be able to send a nuclear warhead weighing much more at the United States at far lower altitude over the Arctic. Meanwhile, many Americans developed instant contempt for their own system, which they suddenly saw as complacent, flatulent, and lacking a serious educational system. If even Slavs could beat the United States to space, after all, the situation was really dire. Or so they thought.

To his credit, Eisenhower the former soldier did not want what was clearly the new frontier of space to be completely militarized. He believed that all governmental organizations, like those in the private sector, are inherently self-aggrandizing. But that trait was uniquely dangerous where highly competitive military establishments were concerned because self-aggrandizement, justification for ever more weapons development and procurement, could lead to war. He was therefore determined that space be militarized only to the extent that the nation's fundamental security was guaranteed. Von Braun's pleas to use a Jupiter missile from the Army Ballistic Missile Agency to answer the Russians were therefore rejected. Instead, the White House gave the go-ahead to a sleek navy scientific rocket launcher and satellite called Vanguard, which was not a ballistic missile, and which looked pretty, not menacing.

Then the Communists struck again, this time with a wallop. On November 3, they orbited a mutt named Laika in *Sputnik 2*. This was alarming for two reasons. The new Sputnik weighed eleven hundred pounds. That approached warhead weight. So the enemy got to test a long-range nuclear ballistic missile and score another propaganda victory with one shot. More ominously, sending a sensor-rigged mammal to space could have only one implication: a human was to follow.

On December 6, 1957, with television cameras trained on it to provide national coverage, and assorted celebrities watching from a safe distance, the navy's beautiful Vanguard rocket blew up on its pad at Cape Canaveral seconds after ignition. Widespread mortification set in even before the grapefruit-sized, chrome satellite rolled clear of the greasy black and orange fireball. With the fourth estate calling the American entry Flopnik, Stayputnik, Kaputnik, and worse, von Braun was finally given permission to carry the nation's colors to space. On January 31, a Jupiter C (prudently named Juno, after Jupiter's peace-loving wife and sister, thereby respecting the president's sensibilities) carried a JPL-developed satellite named *Explorer I* to orbit.

Six months later, Public Law 85-568, the National Aeronautics and Space Act of 1958, authorized the creation of a national space agency to be known as the National Aeronautics and Space Administration. Acceding to Eisenhower's wishes, the framers wrote in the Declaration of Policy and Purpose: "The Congress hereby declares that it is the policy of the United States that activities in space should be devoted to peaceful purposes for the benefit of all mankind." And, they stated explicitly, the welfare and security of the country required aeronautical and space activities. "The Congress further declares that such activities shall be the responsibility of, and shall be directed by, a civilian agency exercising control over aeronautical and space activities sponsored by the United States." The exception related to weapons and military space operations, which would come under the control of the Department of Defense. But "security" was infinitely ambiguous. If a civilian "first" in space was a political triumph, it could be said to enhance national security, especially in the gray area where political and military activities overlapped. Furthermore, the civilian and military programs grew out of a common technology, ballistic rocketry, and shared key personnel. Von Braun and the others on his rocket team were first and foremost missile designers, for example, and the first astronauts would be "fighter jocks."

There was no doubt in 1958 that the military would have a substantial role in space. As early as 1947, the Rand Corporation, a California think tank, saw the eventual need for military reconnaissance satellites. And in June 1958, even before the legislation creating NASA became law, the air force was thinking about manned space planes in an ultimately doomed program that would come to be called Dyna-Soar. Yet Eisenhower's legacy to the space program, which would soon start the exploration of the solar system, was profound. It tried to distinguish between the civilian and military sectors, if only politically, and gave primary responsibility for scientific investigation and a human presence in space to the civilians. The military was assigned the task of protecting national security, which included assuring the safety of the nation's growing commercial interests in space, as the Royal Navy had done for British merchantmen on the high seas during the eighteenth and nineteenth centuries.

John F. Kennedy's legacy to the nation's space program was a superlative moment, followed by a conundrum that haunts NASA to this day. But it was not his legacy alone. By 1960, when the senator from Massachusetts beat Richard M. Nixon for the presidency, the nation was reeling from what appeared to be an indomitable Soviet—read Communist—space program. *Sputnik 1* started the space age. *Sputnik 2* was a clear indicator that the Russians were determined to send men to space. *Sputnik 3*, which went up on May 15, 1958, was solar-powered. Meanwhile, there were three unannounced (and unsuccessful) Soviet attempts to reach the Moon with unmanned spacecraft between May Day, 1958, and the following December. Then, on January 2, 1959, the Russians scored yet another first when *Luna 1* became the first machine from Earth to reach the vicinity of the Moon, which it passed at a distance of thirty-five hundred miles before heading toward the Sun. And while the U.S. Explorer scientific program was ringing up quiet successes, a series of failures in other programs plagued Washington. On March 3, 1959, though, *Pioneer 4* sailed past the Moon at a distant 37,500 miles and then headed for a solar orbit like its Soviet predecessor. There was now no doubt that a space race was under way, that international political prestige accrued to whichever side seemed to be winning (whatever that meant), and that the Union of Soviet Socialist Republics was crowing loudly about one first after another.

It was obvious to just about everyone in the fledgling U.S. program, certainly including von Braun, Oberth, Krafft Ehricke, Eberhard Rees, Ernst Stuhlinger, and other Peenemunde alumni, that the Russians were quietly beginning to aim what they would soon call cosmonauts at the Moon. So as early as December 15, 1958, von Braun and some of his colleagues recommended that NASA start planning to land Americans on the Moon by 1967.

At the same time, congressional concern about the Soviet program led to the creation of the Senate Committee on Aeronautical and Space Sciences, chaired by Senator Lyndon Baines Johnson of Texas, and the House Committee on Science and Astronautics. There was now no doubt that politics, no less than propellant, was fueling U.S. rocketry.

Early in 1960, the House committee heard George Allen, director of the United States Information Agency, testify that the space program had "an importance far beyond the field of activity itself. . . . It bears on almost every aspect of our relations with peoples of other countries and on their view of us compared with the USSR. Our space program may be considered as a measure of our vitality and our ability to compete with a formidable rival, and as a criterion of our ability to maintain technological eminence worthy of emulation by other peoples." The head of the government's international propaganda mill was warning that an also-ran space program could cause the United States to look like a second-rate superpower to the third world, and even to its allies.

The problem as it existed in the last stages of the Eisenhower administration, and as it exists today, was that there were multiple and competing motivations for activity in space. Given the relatively meager space budget, Ike's insistence that civilian and military operations appear to be separate, and his administration's stubborn resistance to a coordinated national space policy that combined civilians and soldiers, rivalry and fragmentation inevitably infected the space program. And to make matters worse, the civilian and military programs' tight budgets discouraged the development of a potentially powerful industrial base. The action for aerospace corporations was therefore in air force and navy ballistic missiles and other big-ticket programs.

It is not entirely clear what motivated Kennedy to call for the landing of Americans on the Moon. But if it can be said that there was one day when that idea and the larger context of international relations coincided, it was April 19, 1961. Two days earlier, an invasion force of CIA-trained Cuban exiles trying to overthrow Fidel Castro had gone ashore at the Bay of Pigs and were quickly cut down as they waded through the surf and onto the beach. On the night of April 18 and into the nineteenth, Kennedy and his closest advisers decided not to support the exiles militarily. The destruction of what all the world knew was a U.S.-sponsored, but ultimately unsupported, military action was humiliating. So, too, had been the flight of *Vostok 1* only a week earlier. It's occupant, a boyish and beaming cosmonaut named Yuri Alekseyevich Gagarin, became the first human in space when he flew one complete orbit around Earth. There was no longer the slightest doubt about why Laika (and other mutts named Belka and Strelka) had been sent to space. Premier Nikita S. Khrushchev, who was record-obsessed and used his country's space program for international political leverage, bragged loudly about the achievement. Kennedy was an intensely competitive Irish-American who hated to lose at anything, let alone at international prestige.

So Kennedy called Johnson into his office a week later and asked him what direction he would recommend for the space program. He also asked Theodore C. Sorensen, his special counsel, and Jerome Wiesner, his science adviser, to review the outlook for the next steps in space. And the space agency itself weighed in with several options, including planning two-man orbital missions, a space station, and sending a manned rocket around the Moon and back. But NASA's leaders also admitted that the prospects for beating the Russians at any of those feats were poor because of their head start in heavy rocketry. JFK was told that the first really good possibility for getting ahead of the opposition would therefore be relatively long term: to have von Braun's rocketeers develop a monster launcher that would send Americans to the Moon.

"The President was more convinced than any of his advisers that a second-rate, second-place space effort was inconsistent with the country's security, with its role as world leader and with the New Frontier spirit of discovery," Sorensen wrote years later. On the basis of all the advice Kennedy got from his inner circle and NASA, he then made what he later called one of the most important decisions of his presidency: "to shift our efforts in space from low to high gear."

On May 25, 1961, in a speech to Congress that addressed "urgent national needs," Kennedy listed several dangers facing the United States, including Communist subversion. He then used the Sputnik and Gagarin flights, and their impact on "the minds of men everywhere," to call for the United States to land a man on the Moon and return him safely to Earth before the end of the decade. "While we cannot guarantee that we shall one day be first," he warned, "we can guarantee that any failure to make this effort will find us last."

With that, in John M. Logsdon's view, Kennedy not only set a single overarching goal for the space program, he fundamentally changed the nature of the program itself. "He challenged the assertion that a 'single civil-military program . . . is unattainable' by approving the initial plan for just such a program, aimed at establishing American preeminence in every aspect of space activity, civilian and military, scientific and commercial, prestige-oriented and unspectacular." He thereby abruptly and dramatically reversed Eisenhower's decentralized and lackluster space program.

And he did far more that that. If being first to land men on the Moon was by definition a race, then the time factor was crucial. That, in turn, meant the most immediately available rockets—in effect, proven, off-the-shelf hardware—would have to be used, at least at the outset. But the existing rockets, really intermediate-range ballistic missiles, could not get men anywhere near the Moon. The John C. Marshall Space Flight Center at Huntsville was therefore mandated to develop within a relatively short period a rocket launcher that would carry astronauts to the lunar surface and bring them home. The tight time constraint eliminated any possibility of designing a conceptually exotic launch system. The proven hardware used to realize that specific goal would have to be a launcher based on the reliable old standby, the ballistic missile. But it would be a monster missile: a three-stage superbooster called Saturn V that would tower 363 feet high, weigh 6.2 million pounds, and produce 7.5 million pounds of thrust in its first stage alone. Basically, it would toss three astronauts in a complicated spacecraft at the Moon in what was unofficially called the cannonball approach, a curious evocation of Verne's colossal Columbiad Moon cannon. So Saturn V was inherently a huge ballistic missile, and the men who would ride it to the Moon were air force and navy pilots, two factors that also reversed Eisenhower's insistence on the separation of the civilian and military sectors. The Russians would soon be headed in the same direction with the development of a disaster-plagued monster rocket called the N1, the development of which was closely monitored by U.S. reconnaissance satellites.

Gone was the development of reusable spacecraft that began in the X-15 program, which was to turn the space around Earth into an ocean plied by people-carrying space planes that could be refueled and refurbished like airliners. Gone was the notion of building a large space station as an embarkation point from which to start a lunar colony and launch an expedition to Mars. Gone was the grandiose plan to gradually extend the human presence steadily outward as envisioned by dreamers from Tsiolkovsky to Goddard to Korolyev to von Braun himself. To this day, the gradual, methodical move to space as it was described in Collier's is called the von Braun paradigm. As the master rocketeer had been co-opted by the German army because of the imperative to extend the range of its artillery, so was he co-opted by Kennedy's fixation on the political imperative of landing men on the Moon before the

Communists did. But for von Braun the space dreamer, that still beat designing weapons by a long shot.

By September 1963, when the two-man Gemini program that would lead to Apollo was well under way, Kennedy had a change of heart that could have profoundly altered the course of the space age and probably the Cold War itself. Having forced Khrushchev to remove ballistic missiles from Cuba in October 1962, and then having come to an agreement with him on the need for an innovative limited nuclear-test-ban treaty, Kennedy began to feel confident enough politically to consider going to the United Nations to suggest ending the space race and turning the Apollo program into a cooperative one with the Soviet Union. That is, he thought sending astronauts and cosmonauts to the Moon together could significantly lessen Cold War tension and signal a new and dramatic move toward peace. He would not have been unmindful that such a magnanimous gesture would further assure his place in history. Kennedy rationalized the idea by noting that the National Aeronautics and Space Act explicitly gave him the power to "engage in a program of international cooperation" for peaceful purposes. But when word of the plan reached Capitol Hill, exasperated House Republicans who were adamant about not sharing what figured to be a monumental American triumph with Communists (not to mention technology), passed an amendment blocking it. It is interesting to speculate on what might have happened had Kennedy not been assassinated two months later and had applied his renowned tenacity to going to the Moon with the opposition instead of in competition with it.

Neil Armstrong and Buzz Aldrin's landing *Eagle* on the Moon on July 20, 1969—within the decade—transfixed the people of Earth. As Carl Sagan wrote with characteristic perception, "The Moon was a metaphor for the unattainable: 'You might as well ask for the Moon,' they used to say. Or 'You can no more do that than fly to the Moon.' " Millions who watched the event on television or heard it described on the radio understood that they were privileged to live at a transcendental moment: when their race first reached another world. And whatever Armstrong said about a small leap for man and a giant leap for mankind, the flag planted on the Sea of Tranquility, which Aldrin made a point of saluting, was American. The message was not subtle and it was wasted on no one: the most daring feat of exploration in human history, and one of the truly outstanding technological and managerial triumphs of all time, was accomplished by American genius, muscle, and money. (Ironically, and largely missed in the excitement and drama of the moment, it also proved Eisenhower had been right about the fundamental strength and resilience of American space science and technology.) And the triumph of Apollo 11 was also a death-defying political stunt of fantastic proportions. With the relentless, horrific explosions of earlier rockets still in their collective mind, Americans watched three of their intrepid young pilots risk their lives to land on another world for the first time, bringing glory to themselves and their country and demonstrating in the most dramatic way that their system was superior to their rival's. It was a time, as someone said later, when anything seemed possible. Pan American World Airways, the nation's preeminent international carrier, was taking bookings for spaceliner trips to the Moon. It was inspired public relations.

But the feat itself, a modern odyssey of immense proportions, created a formidable problem for the space agency. How long could it keep the audience's attention and enthusiastic support after death had been defied in the ultimate high-wire act?

The problem of what to do after Apollo had been anticipated at least five years before that first Moon landing. In 1964, President Johnson told NASA to form a post-Apollo Future Programs Task Force to find a direction for the space program. Meanwhile, NASA administrator James E. Webb made vague references to robotic missions to Mars and somehow using the Apollo-Saturn system in the vicinity of Earth. "Over the next five years vagueness became the pattern," historian Walter A. McDougall has written. "Webb apparently thought it poor strategy to promulgate long-range plans, when the costs of new programs were unpredictable and sure to seem excessive in the Vietnam/Great Society era. Inevitable tinkering with long-range plans might also make NASA appear to be confused or fishing for big money." Meanwhile, the diverse centers themselves, protecting their turf, sharply disagreed on what the space agency's priorities should be, while Webb fought to protect Apollo and the projected follow-up missions. He even desperately resorted to warning darkly about new Soviet threats, and even claiming his agency's management skills and expertise in cutting-edge technology could be applied to the war in Vietnam, community planning, police work, firefighting, transportation, medical care, and to foil airliner hijackings.

But it was no use. Nothing grand was in the works. So on September 16, 1968, Webb announced that he was going to leave NASA on October 7, his sixty-second birthday, to pursue interests in education and urban affairs. NASA insiders, however, knew there was more to Webb's departure than was made public. According to one credible account, Webb urged Johnson to restore various budget cuts and threatened to resign if LBJ did not do so. Lyndon Johnson had a notorious ego and did not like being threatened by anyone, let alone a subordinate. He therefore responded by calling in the White House press corps and announcing that the beleaguered space agency administrator had something important to announce. A stunned James Webb announced that he was leaving.

*Eagle*'s alighting on the Moon and the immense technological competence and political resolve that had propelled it there had been a dramatic epic of Wagnerian proportions. But while NASA's armor-clad heroes successfully defied death for God and country, others in the cast, to continue the metaphor, were wandering off, sleeping, arguing among themselves, scrounging for sustenance, or dreaming of adventures that had no basis in reality.

Von Braun remained foremost among the true believers, though there were many others, right down to the members of the rocket societies. In 1964, *The New Scientist* magazine had published a series of articles in which ninety-nine of the intellectual establishment's leading lights predicted what they thought their disciplines would have accomplished twenty years from then, in the Orwellian year of 1984. Ever the dreamer, von Braun's contribution had been an essay called "Exploration to the Farthest Planets," in which he predicted that people would either have already landed on Mars or at least made a close approach. The same for Venus. The Moon, he continued, would by then have been turned into a manned colony. "Lunar landings will have long since passed from fantastic achievement to routine occurrence. Astronauts," the savant of space added, "will be shuttling back and forth on regular schedules from the earth to a small permanent base of operations on the moon." And sitting on the lunar surface would be an observatory that would do nothing less than discover the nature, extent, and origin of the entire universe.

But that was not to be. For a number of reasons, von Braun's extravagant vision did not materialize. The most important of them was simply that there did not seem to be a compelling reason to leave Earth on expensive and dangerous forays into space when there were immediate and dire needs at home. While those who believed in the manned space program with almost religious conviction loudly celebrated the voyages to the Moon, they remained a small minority. There was no real constituency for more adventures in space. Five polls conducted between 1961 and 1970 showed that, with one exception in October 1965, more Americans thought the government should not fund human trips to the Moon than thought it should. Other polls taken between 1962 and 1972 showed that while between 60 and 80 percent of respondents thought Apollo was worth its cost, only 35 to 40 percent approved of the program. And in 1969, when Apollo 11 and 12 reached the Moon, slightly more than half of those questioned told the pollsters that too much was being spent on space.

Four months after Armstrong, Aldrin, and Collins reached the Moon, the crew of Apollo 12 returned there in the *Yankee Clipper*, with Charles "Pete" Conrad and Richard Gordon spending almost eight hours on the Ocean of Storms and returning with seventy-five pounds of rocks.

But by the time Conrad, Gordon, and Alan Bean splashed down in the Pacific on November 24, a perceptible disinterest was taking hold in Washington and beyond. The novelty was wearing off. It didn't take a P. T. Barnum to anticipate this. In February 1969, even before the flight of Apollo 11, President Nixon had ordered that a Space Task Group be formed to chart the space program's course after the Apollo program had ended. The STG was chaired by Spiro T. Agnew, the vice president, but otherwise contained such heavy hitters as Secretary of the Air Force Robert C. Seamans, NASA administrator Thomas O. Paine, and Lee A. DuBridge, the president's science adviser. Their report echoed one written by the President's Science Advisory Committee two and a half years earlier. It opened by noting, "The manned flight program permits vicarious participation by the man-in-the-street in exciting, challenging, and *dangerous* activity [italics added]. Sustained high interest, judged in the light of current experience, however, is related to availability of new tasks and new mission activity—new challenges for man in space." This reflected an implicit but usually unspoken belief that dangerous activities were not only good, but necessary, and that increasingly spectacular feats would have to be staged to capture taxpayers' attention. There followed a Chinese menu of options that included sending Americans to Mars within fifteen years while pushing unmanned solar-system exploration, enhancing national security, and contributing to the quality of life on Earth. All of them were to be done on the cheap.

Similarly, the President's Science Advisory Committee's own report, *The Space Program in the Post-Apollo Period*, called for a "balanced program" with five major objectives: a limited extension of Apollo to continue exploring the Moon; a "strongly upgraded" program of unmanned exploration of the solar system; development of technology for long-duration manned planetary exploration; the exploitation of space resources for national security and the social and economic well-being of the nation; and a vague call for the advancement of science.

Both reports provided a broad spectrum of programs that were compromises reflecting many diverse, competitive, and politically active, though relatively small, civilian constituencies. (The military, and especially the air force, was yet another hungry constituency.) The conflict between proponents of manned flight, versus purely unmanned scientific missions, had begun before the space age started. And there was, in turn, constant funding battles within those two camps themselves over specific programs. The lack of focus on a single, unifying objective beyond the Apollo program, which was targeted for termination even before the first astronauts reached the Moon, forced NASA to try to be the proverbial jack-of-all-trades.

By April 1970, with the first two lunar landings and the routing of the commies in space accomplished facts, a palpable ennui set in across the country. One NASA history publication later noted with impressive honesty that by late December 1972, when Apollo 17's Gene Cernan became the last man on the Moon, there was increasing concern in Washington over what some took to be the space agency's bloated budget. And, the history went on. "Even the general public, if not quite bored with it all, was taking 'moon shots' in stride."

More fundamentally, big science, which had won World War II and decisively helped end the scourges of Nazism and Japan's insatiable imperial ambitions, was now taken to be a mysterious Strangelovian octopus. It was seen by the left as providing the armed forces with the means to massacre "liberation" armies in Southeast Asia and elsewhere and creating fearsome weapons that could end the world.

And if that weren't bad enough, big science was also thought to be devouring resources that were needed to ease social problems and address health issues. Atomic physics, for example, consumed vast amounts of funds to pursue abstract theories, which almost no one understood, and develop horrific weapons, which almost everyone feared.

So by that spring of 1970, most of the public had begun to wander away from the Moon landing program, while a growing number of intellectuals and academics attacked it as a "moondoggle" and condemned it for being irresponsibly wasteful. Left-wing intellectuals, who had always been hostile to science, in part because its basis is a system of rigid proof that is by definition antithetical to their often reflexive, slapdash polemics, found in every aspect of the space program a high-visibility target of opportunity. Astronomers who wanted to study the rings of Saturn, physicists who were eager to use the interactions of Jupiter and its satellites to model the solar system, and geologists who looked to the Moon for clues about how Earth was formed were derided as stealing vast sums from the poor, the ill, and the socially dispossessed for their own narrow, irrelevant, and selfish purposes.

But at least the physicists and other scientists who worked in laboratories were for the most part unseen and therefore anonymous. The whole space program, by contrast, literally had the highest profile of any large-scale enterprise. It was therefore the easiest target. NASA's renowned public relations apparatus and the space-agency culture it reflected, which was mandated to sell the program to the public, depicted astronauts as bright and brave individuals who had no personal shortcomings and who were pioneers in the settlement of space. A careful balance was struck between insinuating that manned space operations were dangerous—the Agnew-committee paradigm—and promoting space as a natural place for humans to live and thrive; that people belonged there. The deaths of Virgil I. Grissom, Edward H. White, and Roger Chaffee in a flash fire in their command module on January 27, 1967, and Apollo 13's disastrous flight to the Moon three years later, were proof enough that there were life-threatening risks in taking to space. At the same time, television images of astronauts apparently so at home on the "final frontier" that they could cruise around the lunar surface in battery-powered lunar-roving vehicles as though they were out for a Sunday drive, hit a golf ball, and hop like inflated rabbits made living on the Moon and elsewhere in space seem relatively easy and in the scheme of things.

The intensity of the drive to reach the lunar surface within the time limit mandated by Kennedy, the pervasive publicity given to the Mercury and Gemini programs that led to it, and the sheer grandeur of the enterprise made reaching the Moon appear to *be* the space program. That, too, was pernicious.

To be sure, there was activity elsewhere. Venus and Mars had been reconnoitered by Mariner spacecraft long before Apollo 11. A slew of scientific, weather, navigation, and communication satellites had been sent into orbit around Earth and made revolutionary contributions to civilization. And in 1968 another Pioneer space probe went into a scientifically useful orbit around the Sun. Two months before Apollo 15 carried the first rover to the Moon, the Jet Propulsion Laboratory's Mariner 9 became the first machine from Earth to orbit another planet and sent home some seven thousand detailed pictures of the Martian terrain that enraptured the mission's scientists. But where most of the public was concerned—Joe Six-Pack, as it was later collectively (and somewhat derisively) called—these missions were irrelevant and therefore absolutely unimportant. Science was taken to be incomprehensible and inconsequential by the vast majority of people, including many with higher educations. And so, because these other missions were taken to be peripheral compared to the great human adventure that sending out astronauts represented, they were for the most part ignored. The long-distance robotic missions were occasionally tweaked by competing programs even within the space agency itself. NASA assistant administrator Franklin D. Martin would give a succinct explanation for the primacy of sending humans to space at a meeting in Washington twenty years after the *Eagle* landed: "We don't give ticker-tape parades for robots," he quipped.

But once the overwhelming goal of getting Americans to the Moon before Russians reached it had been accomplished, once the great race had been won, the stands in effect cleared and the cheering spectators wandered off. And a large chunk of the gate receipts went with them. The space agency's overall budget went into a long decline starting in 1966, when most of Apollo's research and development work had been completed and much of the equipment already bought. The budget specifically for manned spaceflight shrank steadily from 1969 to 1974, when it dropped below $1 billion for the first time since 1962. Apollo alone had cost $25 billion, almost all of it spent to get the *Eagle* onto the Sea of Tranquility. With a projected budget that would shrink after Apollo 11, and a perception that public interest would dramatically lessen after the Moon had been reached, it was quietly decided to kill Apollo 18, 19, and 20, the last three of the planned missions to the Moon.

Now NASA was out of focus and disarray set in. The comprehensive and orderly plan for moving to space envisioned at the Hayden Planetarium meeting in 1951 and recorded on the pages of *Collier's* had fallen victim to the Moon landing program's outstanding success. And that program was so expensive it caused deep cuts elsewhere in NASA that continued to thwart following the old master plan. Everyone in the space business remembered the plan, of course, but a leaner NASA was now forced to try to pursue it piecemeal.

It is a measure of how severely NASA was crippled that the instrument for trying to pursue the old plan was called Apollo Applications, a program that was simply supposed to make good use of what remained of Apollo's launch vehicles and other hardware. And with the scarcity of funding being what it was, even the applications people had to do was difficult while being second-guessed and micromanaged. Harold B. Finger, NASA's associate administrator for organization and management, deplored that his agency had gone from being "a large, fast-moving program aimed at a clearly identified national objective to a situation in which reductions have been made in our program and, therefore, to a requirement for a very detailed examination of every element of our program before final approval is given to proceed."

The result of that tortuous process was the creation of the first successful space station. Although von Braun was intimately involved in its development, it was anything but the grandiose rotating wheel he had originally conceived, since designing a station from scratch, and particularly a big one that had to be assembled in orbit, was out of the question. The station was called *Skylab*, and it was a paradigm for what ailed NASA, fiscally and technologically. In 1965, under the aegis of the Apollo Applications Office, engineers at Marshall began studying the feasibility of turning a Saturn IVB into a space laboratory, called an orbital workshop. The IVB was the third stage of the massive Saturn 5 launcher. This forty-eight-foot-long, twenty-one-foot-wide cylinder weighed a little more than eighty-two tons and had been designed to give the Apollo spacecraft their final push. Apollo Applications came up with two ways to convert IVBs to stations: a "wet" approach in which astronauts made them habitable after their fuel was burned up, and a "dry" technique in which they would be launched as laboratories and would not therefore be filled with fuel. By 1966, NASA was planning to launch three of the hybrid wet stations and three dry ones. The number was gradually whittled down until only one dry station remained. It was launched on May 14, 1973, immediately after a series of frantic and disastrous attempts by the Russians to get their own station into orbit first. But its American counterpart suffered a serious accident right at the start. The meteoroid shield, which was supposed to ward off speeding rocks, did not open properly. It jammed one of two solar panels and tore off the other.

Eleven days after *Skylab* was launched, it was joined by a three-man crew on a Saturn IB, which was a smaller version of the Saturn V designed to get crews to Earth orbit but not all the way to the Moon. The astronauts stayed on board for twenty-eight days and were followed by a second crew that lived on *Skylab* for two months. They were in turn replaced by a third and final crew that stayed for a record-breaking eighty-four days. The primary purpose of keeping nine men in orbit for 172 days was to see how they reacted physiologically and mentally to long stays in the near weightlessness of low Earth orbit. In that regard, *Skylab* won high grades for the quality of its science return, especially as a precursor to longer manned missions. But that was precisely the problem. The mission made sense only if the collected data were applied to longer-duration flights, one being on a permanently manned station, and another on a round-trip to Mars. Assembly of a station of sorts would tentatively be started at the end of 1998. The Mars mission, or even a return to the Moon, were definitely not on the agenda.

*Skylab*, which was becoming difficult to control, was brought down on July 11, 1979. It hit the Indian Ocean blazing like a meteoroid. On July 15, 1979, two cosmonauts in a Soyuz capsule and three astronauts in the last remaining Apollo capsule were sent into orbit in a political stunt dreamed up by Secretary of State Henry Kissinger to show that détente between the U.S. and the USSR even extended to space. It was called the Apollo-Soyuz Test Project. The two spacecraft docked on July 17, and for the next two days, smiling Russians and Americans made a show of entering each other's capsules, shaking hands, exchanging souvenirs, and exuding the camaraderie of détente. The Soyuz parachuted onto the Russian steppes on the twenty-first. The Americans came down in the Pacific three days later. They were the last astronauts to go to orbit until a fundamental element of the original master plan, which had been approved three years earlier by Nixon, was launched. It was a make-do version of von Braun's shuttle. Meanwhile, the carcasses of the two remaining Saturn V's were laid on their sides at the entrances to the Kennedy Space Center at Cape Canaveral and the Johnson Space Center outside Houston. The manned space program would be grounded for six years. Although few understood it at the time, the first space age was over.

Manned spaceflight as it began and was developed during the Cold War was the child of political and military competition between the superpowers and was sold to the world as a great and inevitable adventure having enormous scientific merit and enduring political prestige. It had purpose and was focused.

As much could not have been said about the space shuttle. On September 1, 1969, with missions to the Moon barely under way and the Apollo program understood to be finite, von Braun's old dream of a fleet of reusable space transports was dusted off. NASA hatched what it termed a Space Transportation System that consisted of two interlocking parts: shuttles and a station. STS was cleverly conceived because either part was dependent on the other. Aside from carrying all manner of satellites and other spacecraft to low orbit, the shuttles were to be used to cobble together a prefabricated station. Building a station was therefore one of the shuttle's most important jobs. At the same time, the station would be wholly dependent on shuttles for crew changes and supplies. NASA, going back to the original script, therefore proposed the combined Space Transportation System as the next logical step in what was still called the conquest of space.

That day, September 1, 1969, Nixon's Space Task Group recommended funding a shuttle program that would be operational by 1975–77. Funding for the other half of the system, the station, was not approved. In its report to Nixon, the Space Task Group naïvely maintained that the shuttle would "carry passengers, supplies, rocket fuel, other spacecraft, equipment, or additional rocket stages to and from orbit on a routine aircraft-like basis," with emphasis on commonality of parts and missions, reusability, and economy. The idea, logical on the face of it, was that using the same machine repeatedly would be cheaper than using so-called expendables, which were the traditional rocket boosters that dropped into the Atlantic and disappeared after launch. As every airline executive knows, the more passengers and freight an airliner carries, the cheaper it is to use. So with the shuttle. "The greater the number of flights, the greater the advantage to a reusable system," a General Accounting Office report on the shuttle explained. Accordingly, NASA assured the GAO in April 1973 that there would be 779 flights during the first thirteen years of operations, or sixty flights a year, or an average of five a month, or one every six days. That was patently absurd. But the space agency was desperate for a follow-on to Apollo so it resorted to wild exaggeration. The stark reality, as the bureaucrats in NASA saw it, was that they would be marginalized, if not virtually closed down, without a big core engineering program to take the place of Apollo. And the shuttle program was also a profitable undertaking for the aerospace industry, which was running on dry in the wake of Apollo. Exaggeration for what the fraternity implicitly believed was an eminently worthwhile cause could therefore be justified.

And if the rationale for using shuttles was their flight frequency, it followed they had to be the only game in town, so the expendable program was killed outright. In addition, the search for shuttle customers was broadened to include a reluctant air force, which was signed on, at least temporarily. The shuttle orbiter's payload bay measures fifteen feet by sixty feet to accommodate the supersecret KH-11 reconnaissance satellite, then code-named Kennan, which was the size of a bus. The air force soon insisted that spacecraft, including reconnaissance satellites, that had to be launched quickly during crises could not be dependent on shuttles. It successfully made its case and was allowed to continue using large Titan expendables.

That having been said, it would be unfair to describe the selling of the shuttle in wholly cynical terms. The men and women in the program, whether in government or industry, tended to be wholehearted believers that it is the human race's destiny to take to space. Francis Clauser, for example, was anything but a starry-eyed dreamer ungrounded in the relevant technology. He was chairman of the College of Engineering at Caltech and a member of the panel that reviewed the space program after Nixon's election. In May 1969, Clauser predicted that the decade ahead would "see the cost of space transportation reduced to the point that the average citizen can afford a trip to the Moon." He went on to explain, "When I speak of *low-cost* space transportation, I define low to be so low that the *citizenry* can afford to buy tickets for space." Lockheed's Max Hunter told an audience at the University of Michigan in mid-1970 that ninety-five shuttle flights a year would bring the cost of a flight to $350,000, or $7 a pound, instead of the $1,000 a pound it cost at the time. At $5 a pound, he predicted, the Hiltons would build an orbiting hotel. And NASA associate administrator George E. Mueller said at a shuttle symposium in Washington in October 1969 that it was feasible to lower the cost of orbiting a pound to between $20 and $50. Doing so, he added, would "open up a whole new era of space exploration." They absolutely believed what they said. And a chorus of other highly informed individuals believed it, too.

Nixon formally approved the shuttle on January 5, 1972. As originally conceived by NASA, the shuttle was to consist of two vehicles: a huge winged launcher that would take off on its own power and carry on its back a smaller delta-winged orbiter that contained a payload of people and cargo. After the orbiter was released and heading skyward, the launcher would swoop back to Earth and land like an airplane, to be refurbished and fitted with another orbiter. But the concept struck the Office of Management and Budget as unnecessarily expensive. OMB therefore ruled that the shuttle could cost no more than $5.15 billion to develop. That killed the winged, reusable launcher. NASA instead had to settle for an orbiter that would be attached to a huge external fuel tank that fed its main engines and then dropped off and fell into the sea. It would also use two reusable solid rocket boosters that would help launch the STS straight up.

After a series of engineering setbacks, the two most notable being trouble with the orbiter's main engines and keeping its heat-resistant ceramic tiles glued in place, the shuttle *Columbia* roared away from the Kennedy Space Center on April 12, 1981. It was anything but coincidental that it was twenty years to the day after Yuri Gagarin's flight. Both the American and Soviet programs had a mischievous penchant for poking the other in the eye whenever possible.

Five months earlier, perhaps to convince the public that the program was more under control than it really was, NASA had issued the Space Transportation

System's first five-year timetable and payload manifest. It called for thirty-seven flights between August 31, 1981, and May 14, 1985. It also predicted that there would be five hundred flights by the end of 1991 (a quiet reduction from 779).

The shuttles forced a fundamental change in promotional strategy on an agency well aware that the glory of Apollo was fleeting and that it was not on the public's short list of federal institutions providing indispensable services. The old derring-do approach was abandoned and replaced by an image of the shuttle program as an integral and beneficial part of society and one that would make people at home in space, where they belonged. The operative word was *relevant* to people's lives and society's needs. This variously came in the form of socially important "spin-offs," such as computers, whose development, or at least accelerated development, was credited to the shuttle program, and in the health area. Experiments carried in the payload bays of the four orbiters were touted as possible landmarks on the way to curing cancer, radiation poisoning, and AIDS, among other scourges. Of course they were possible. But they were far from likely. *Possible, may,* and *could* were used by the space agency and contractor publicists to portray the shuttle as an integral part of medical and health research. A picture of a seriously ill infant was run in newspapers with an implicit explanation that the shuttle program could save it. One major contractor ran an ad showing an older child crediting the program for her father's job. Money for space, the saying went, was spent on Earth.

That the orbiters were large enough for their occupants to work in flight suits and shirtsleeves moved the user-friendly space-environment image far beyond hitting a golf ball in a bulky space suit on the Moon. Early flights showed delighted astronauts chasing blobs of water to drink, watching eating utensils and other implements floating in midair, and they, themselves, tumbling, rolling, and standing upside down in near-zero gravity. Plans were eventually made to send a schoolteacher up to teach a lesson from space, and then a journalist to describe the experience to the world. Spacelab, a specially designed science module carried in the payload bay, was hailed for its experiment package after it was first launched in December 1983, and payload space for so-called Getaway Special science experiments for college students and others was set aside for as little as $3,000 each. The unwavering plan was to institutionalize the Space Transportation System until it became a kind of user-friendly superairline; an integral part of the nation's conscience and infrastructure.

But that, too, had an inherent, unavoidable downside. The sight of men and women laughing and frolicking 150 miles in space on a flight that cost $250 million or more did not amuse many in academe, including many scientists, who saw the missions as an aerospace-industry entitlement and frivolous entertainment that sapped funds for serious science, big and small. Worse, the predictability and the apparent ease and safety of successive missions began to be taken for granted by ordinary people, so they in effect stopped looking up. The Space Transportation System was indeed running like an airline (though on a schedule that was so precarious, a budget that was so soaked in red ink, and a frequency so short of the estimated number of flights that it would have bankrupted an airline). The problem that haunted NASA, articulated or not, was that people hadn't been going to airports to watch airliners take off since the 1940s because the wonder of commercial flight had long since worn off.

After a while, journalists and others who watched the space program carefully began calling NASA to task for grossly exaggerating by the hundreds the number of shuttle missions. John Noble Wilford of *The New York Times* quoted one embarrassed NASA official, L. Michael Weeks, as saying, "Every now and then I go back and look at those early projections and have to close my eyes and shake my head."

It was no wonder the discrepancy between projected flights and actual ones was so large. Even if it had been logistically possibly to launch a flight a week, the requirement—the cargo—wasn't there. Some of the missing payloads were to have been the parts of the other half of the STS: the station. But in the absence of a clearly articulated reason to have a station, and with its going through one design change after another, the space station was effectively grounded in a political quagmire.

Then, on Tuesday, January 28, 1986, *Challenger* blew up seventy-three seconds after liftoff, killing all seven on board, including S. Christa McAuliffe, a New Hampshire schoolteacher. The sight of the fiery white flower that exploded in the sky, and of McAuliffe's stunned parents watching their daughter and the others die, traumatized the nation. The shuttles were grounded for thirty-two months while the cause of the tragedy, the faulty O-rings in the boosters, was established and corrected.

By the time the orbiter *Discovery* resumed operations on September 29, 1988, carrying a crew of five and a tracking and data-relay satellite (which could have gone on an expendable) to orbit, the Space Transportation System as originally envisioned had gained its other component, at least on paper. On the Fourth of July, 1982, President Ronald Reagan had called for a permanent

American presence in space in remarks made at Edwards Air Force Base when he welcomed the crew of *Columbia* back from the fourth shuttle mission. The idea was formalized on January 25, 1984, when Reagan announced in his State of the Union message that he was directing NASA to develop a permanently manned space station and, echoing Kennedy, "to do it within a decade." The plan was to have it in orbit by 1992 to honor the five hundredth anniversary of the Columbus voyage. He had made the decision after being sold on the station by NASA administrator James M. Beggs and some administration officials at a meeting in the Cabinet Room the previous month. The only sour note seems to have come from Dr. George A. Keyworth II, Reagan's science adviser, who staunchly opposed the station. Keyworth reasoned that since the Russians already had a station named *Salyut* in orbit, NASA's entry would merely come in second yet again. Instead, he continued, the United States should "leapfrog" the Russians decisively by adopting the longer-range goal of establishing a permanent base on the Moon. But it was too long a leap for fiscal conservatives who were fixated on problems closer to home, the chief one being the military threat from the Soviet Union, which Reagan famously called the Evil Empire. So Keyworth was ignored. The president not only approved the station, but ordered that it be named *Freedom* as an in-your-face gesture to the Russians.

In the most honest of all possible worlds, the station would have been named *Quagmire* for the confusion, ambiguity, and political maneuvering it started. As early as three years before Reagan approved the station, there was deep disagreement even within NASA as to what it would be. The Johnson Space Center wanted a large and expensive "space operations center" that would be permanently manned and serve many functions. The Marshall Space Flight Center, on the other hand, wanted "evolutionary" space "platforms" that would be unmanned but man-tended, meaning they would be visited by astronauts. Eventually, the Marshall contingent conceded that the station or stations would be permanently manned as needed.

That disagreement was only one of the first in a series of them that would go on for nearly two decades. Keyworth was by no means the only scientist who opposed the station. Most scientists were in line behind him. Dr. William D. Carey, the executive officer of the prestigious American Association for the Advancement of Science, and Dr. John H. Gibbons of the Congressional Office of Technology Assessment, testified before the Senate Budget Committee that the station would eat up scientific research funding and that, in any case, its scientific value would be negligible. Dr. Edward Teller, the so-called father of the hydrogen bomb, came out against the station because he agreed with Keyworth: it was more important to build a lunar base than settle for an intermediate step. True to form, he warned that if the Russians built a lunar base first, it would amount to a stunning propaganda defeat at best and an unspecified military threat at worst.

The Department of Defense opposed the station because the space pie was finite and an expensive civilian program would inevitably bleed defense-related programs. (And there was a lingering grudge because an early air force space station, called the Manned Orbiting Laboratory, had been killed in the midsixties so Apollo could live.) The fiscally fixated Office of Management and Budget, which under David Stockman had tried to shut down NASA's solar-system exploration program in 1981, said it was voting against the station because it was quite simply unaffordable.

At the heart of the war over the station, beneath the partisan bickering, were three problems: it was expensive; it lacked a clearly defined purpose; and the public was largely apathetic. It didn't take a Nobel laureate in economics to understand that NASA's institutional strength, as defined by funding, lay not in examining Saturn's rings and finding new moons of Jupiter, but in major engineering programs that kept a handful of people in space and many thousands working on Earth to make sure they stayed there. The shuttle and the station were therefore interlocking, mutually dependent, big-ticket programs that largely defined the space agency. But the basic flaw in the process was that it was exactly backward. Rather than defining a clearly articulated goal and then developing the machines to realize it, the machines themselves became the goal and then the search started for ways to justify them.

*The New York Times*, standing back and taking in the partisan fighting and pervasive confusion ten months after Reagan approved *Freedom*, ran a lead editorial in November 1984 called "The Wrong Stuff." It questioned what a manned station that was estimated to cost $8 billion could accomplish that could not be done more cheaply by shuttles and an unmanned station. "NASA has no convincing answers, perhaps because the reason for pushing the space station is primarily bureaucratic. It's a big-ticket, make-work program to keep the agency busy after the shuttle. And it has to be a manned space station because NASA deems human presence essential to retain public support."

Six years later, after several politically inspired design changes, including downsizing it from a dual-keel to a single-keel spacecraft, cutting its power supply in half, and incessant tinkering that produced a slew of other changes, including a nuclear-versus-solar-power debate, its price tag had soared to $37 billion. It was radically different from the smaller version approved by Reagan. And it was still drawing fire. *Science*, the weekly general science magazine, published a commentary quoting several experts as saying that the station was basically a poorly engineered turkey that would require twenty-eight separate shuttle missions to get to orbit and hundreds of space walks a year to keep in operation, and that it was draining more useful programs. With congressional committees, the White House, the aerospace industry, and the space agency all maneuvering for political advantage, and a citizenry that was essentially indifferent to it, *Freedom* became hopelessly stalled.

The contractors distributed identical flyers with only the name of the state changed ("California: Support Space Station") that were passed out at one large annual science meeting in the mid-1990s. They advised people to write or phone their senators and representative urging them to save the space station budget (every senator and representative was listed with his or her telephone number). At stake, the message said, were seventy-five thousand to one hundred thousand jobs, the nation's "economic competitiveness," and children literate in math, science, and engineering. The alleged benefits included enhanced health care ("Space-derived technologies have led to the development of insulin pumps . . . laser heart surgery . . . diagnostic imaging systems . . . reading machines for the blind . . ."), as well as environmental monitoring and the by then familiar commercial spin-offs.

By 1997, with the end of the Cold War and with the participation of Russia, Japan, Canada, and the European Space Agency, the name *Freedom* gave way to the *International Space Station*. On November 20, 1998, after budget overruns that pushed the program's projected cost to more than $100 billion, a Russian Proton rocket thundered off the Tyuratam cosmodrome in Kazakhstan and carried a twenty-ton pressurized power and propulsion module to a 240-mile-high orbit to become the first piece of the *ISS*. The shuttle *Endeavour* followed two weeks later with a hub module that would be the main connector for the modules to follow. Astronauts began connecting the two modules on December 4. Two and a half years later, with astronauts and cosmonauts having lived in the station, and with the Russian Space Agency bankrupt and making its contribution with laundered American funds it was paid for carrying astronauts in its own *Mir* space station, the second Bush White House angered the other station partners by unilaterally eliminating a science module and reducing the station's crew size from six or seven to a maximum of three. The move prompted yet another cry of anguish from the scientific community. "If I were in the White House," one former aerospace industry executive who was on NASA's Advisory Council said, "I would take this as a recommendation to terminate the existing space station."

By then, the old dream had faded with the ghosts of Tsiolkovsky, God-dard, Oberth, von Braun, and the others. The U.S. space program was in vir-tual shambles.

There was a financial guerrilla war by interests—civilian versus military, manned versus unmanned—that were forced to compete for resources made scarce by a society whose power brokers thought most activities in space, cer-tainly including sending people there, were for the most part irrelevant. No fundamentally important reason to justify the space program's existence and growth was given. There was therefore no constituency except the dreamers in the societies and a fraction of the aerospace industry, and they weren't go-ing to claim a serious part of a budget that already had more demands on it than could be accommodated. There was an important reason, of course. But a succession of NASA administrators did not dare publicly define that reason for fear of losing what they already had to expedient politicians whose fa-mously short attention spans extended only to the more immediate needs of their constituents and the next election. The space program became the victim of its own institutional insecurity, myopia, and penchant for tribal warfare.

Early on Saturday morning, February 1, 2003, the space agency and the rest of the country awoke to learn that a torpid situation had yet again turned tragic. Television viewers watched in horror as the orbiter *Columbia* disintegrated in white chunks of debris against a clear blue Texas sky as it streaked back into the atmosphere at the end of a sixteen-day mission. As it approached an altitude of forty miles on its way to a landing at the Kennedy Space Center, superheated air got into a section of its left wing's leading edge that had been damaged at launch and melted part of the wing's inner aluminum structure. That quickly worsened the tear. Since *Columbia* was flying at more than ten thousand miles an hour, the jagged "breach" threw it out of control and it disintegrated into many thousands of pieces. Seven more astronauts, who spent the last minute of their lives knowing they were doomed to die in a spacecraft that was shaking with increasing violence, were killed.

For the second time in seventeen years, shuttles were grounded following a fatal accident, and an investigation was begun. Unlike the grounding that fol-lowed the *Challenger* disaster, however, this time two Americans and a Russian orbited overhead in a station that still required some assembly. The irony that quietly cast a pall over the United States was that the occupants of the *International Space Station* were brought home three months later by a Russian Soyuz spacecraft. And the Soyuz itself missed its landing zone by several hundred miles because of a computer glitch.

The loss of *Columbia* restarted the old debate over the shuttles' value as nearly gravity-free science laboratories. Experiments with tiny flame balls that could allegedly produce more efficient and cleaner combustion engines, and with three kinds of insects, were conducted on the stricken orbiter. Some scientists told journalists that, while they did not want to abandon human spaceflight, they did think it was a wasteful way to experiment. "In order to spend that money and take the risk of a national tragedy if lives are lost, there has to be a very powerful justification," Dr. Gary S. Settles, a professor of mechanical engineering at Penn State, was quoted as saying. "And that justification hasn't been proved." Even Dr. Paul D. Ronney, who teaches aerospace and mechanical engineering at the University of Southern California, and who had the flame ball experiment on board, expressed doubt about using people to haul science experiments:

"If you're looking strictly for science bang for the buck, then I think you would just have unmanned vehicles."

It would come out in the months that followed the loss of *Columbia* that the technological cause of the destruction of the orbiter was only the most apparent part of a far deeper problem. That problem was institutional: the organization had long since lost its way and was languishing in purposelessness. Five weeks after the accident, a NASA official named Henry McDonald, who had led a study team that produced a report three years earlier faulting the space agency for inadequately funding the shuttles and for relying too much on outside contractors, told an investigating board that safety problems were not reliably identified because of poor record-keeping. He explained to the Columbia Accident Investigation Board that the shuttles' managers were responsible professionals, but they were prevented from anticipating serious trouble because of old records that could not be searched with modern tools such as Web browsers to spot potentially dangerous trends. McDonald told the panel that his group's report had urged NASA to modernize its records, but it had not done so.

That was relatively trivial. As the investigation board heard a succession of witnesses and pondered what they said during the spring and early summer of 2003, all kinds of problems began to surface. Some were technical. One expert on aircraft aging, for example, told the board that corrosion and metal fatigue could have played a part in the accident. An earlier Rand study warned that the agency was not paying close enough attention to the fatigue problem. NASA, on the defensive, responded by promising to create an Engineering and Safety Center that could stop shuttle launches at the barest hint of trouble.

Then there was the sorry management situation. The space agency was charged with having institutional communication problems in system evaluation and tracking potential trouble. Even more damning, it was called complacent about safety to the point where inspections declined over time and some inspectors were actually prevented from making spot checks.

All of that was true. But it missed a decisive underlying problem that permeated much of NASA, certainly including the combined Space Transportation System. There was complacency because the manned program had long since become desultory. Three decades of operations that had no overarching goal, no clear focus, no specific purpose, had created institutional boredom and lethargy. Like an individual who lives day to day without a larger reason for existing, the space agency had gradually lapsed into somnolence. With no majestic purpose, no sense of the intrinsic importance of its mission, NASA had become drowsy. That is extraordinarily dangerous for an entity that sends people to space.

The day Henry McDonald testified to the board—March 6, 2003—Sean O'Keefe, the space agency's administrator, shared his own concern with the Senate Governmental Affairs Subcommittee. O'Keefe indirectly referred to institutional boredom when he talked about NASA's retention rate. He warned the senators that alarming trends in retirements, competition by industry for skilled workers, and a decline in graduates with technical skills would severely cut into NASA's skilled workforce. A bill to make the space agency's salaries and benefits more competitive with the private sector's was being debated at the time. So O'Keefe was pointedly asked whether his organization had what it needed to oversee the contractors. He answered by saying that NASA appeared to have the personnel to do the job. Yet he remained worried about the agency hemorrhaging experienced and dedicated people. "What scares me," he added ominously, "is that once they move on, there ain't a whole lot there behind them." Many are moving on, not only for higher wages and enhanced benefits, but because they prefer to abandon an agency that has effectively lost its overall sense of purpose and instead apply their energy and creativity to important and highly focused projects.

Not all of the space program is in the doldrums. While the glory days of

Pioneer, Voyager, Viking, and many of the other trailblazing planetary explorers are long gone, the Galileo, Cassini, and Ulysses missions to Jupiter, Saturn, and the Sun, respectively, were challenging and immensely productive. So were the successively improved rovers that were sent to Mars and which sent home sensational picture postcards that hinted of ancient oceans that could have harbored life. And so, too, were the awesome discoveries by the Hubble Space Telescope and other orbiting great observatories. The missions that study this planet in minute detail, virtually continuously, are fundamentally important. The shuttles or their replacements, and the station itself, are potentially valuable. Yet nothing unites these programs. Nothing gives NASA's scientists and engineers, its managers and technicians, bureaucrats and others, a shared sense of an overriding, common purpose.

But there is a profoundly important mission for the space program. The inhabitants of this planet for the first time have the capability to assure their survival against the vagaries of nature and their own environmentally destructive and treacherous impulses. Ray Erikson, a visionary aerospace engineer in Massachusetts, put it succinctly when he said, "Humankind must wage a perpetual war with Nature, chock full of measures and countermeasures, just to survive. Nature thinks nothing of snuffing several thousand people out of existence with earthquakes, tidal waves, and volcanic eruptions annually." He noted that "house-sized objects (small asteroids, or large meteors) cause Hiroshima-sized blasts in the upper atmosphere *every year.*" (There are roughly thirty such explosions annually.) In that context, Erikson went on, the *International Space Station* "is less a sandbox for scientists than a solidification of the beachhead on the shores of the Cosmos established by the Russian space station *Mir* in 1986." Space, he argued, allows the development of technologies that are necessary to protect Earth from impacts. "With the *ISS*, we have just begun the work needed to establish a permanent, self-sustaining human presence off Earth. This is something we desperately need . . . for our collective continued existence."

That is hardly frivolous. Access to space provides an infinite environment in which for the first time Earth can be protected and enhanced in many ways. Assuring the survival of civilization and the fragile planet that nourishes it is supremely important. It is so important, in fact, that it warrants being the overarching, clearly focused goal of the U.S. space program and its counterparts around the world.

Freeman Dyson has said there are basically two NASAs. There is the real one that operates all the machines, including the shuttles and the station, and which is very conservative. And there is a paper NASA that comes up with grand, adventurous projects it can't afford. Nautical metaphors were a staple of space travel before the space age began. Spaceports, spaceships, and their skippers were long described in science fiction. So were rocket ships. Far-ranging *spacecraft* have been called Mariner and Magellan, Viking, Voyager, and Ulysses. Rescue vessels for space stations are commonly called lifeboats. The shuttle orbiters were named for famous ships: *Enterprise, Endeavour, Atlantis, Columbia, Discovery*, and *Challenger.* Even the diminutive, ill-fated robot explorer Great Britain sent to Mars in 2003 was named *Beagle 2*, after Charles Darwin's ship. Now the metaphor can be applied to the overall state of the space program itself: becalmed.

Knowing this—that the space program, if indeed it can be called a program, has no coherent direction—and acknowledging that frustration with manned spaceflight ranged from moderate to severe, NASA convened a space policy workshop at Caltech on December 10, 2003. Its purpose was to find a meaningful goal or set of them that the space agency could focus on that is realistic and that will engage the public. The two dozen participants included Sean O'Keefe, the NASA administrator, and came from widely varying backgrounds that was supposed to stimulate fresh ideas: to get out of the box, as the current cliché has it. The group was congenial and included space scientists, astronomers, two space historians, a husband-and-wife science-fiction writing team involved in *Star Trek*, and a documentary filmmaker.

Although there was no consensus on what direction NASA should take, there was basic agreement that exploration is a worthy goal for focusing space policy, and that the search for life beyond Earth is an important long-range objective, as are the search for habitable environments and Earth-like planets. The exploration of space by humans is central to what makes the space program interesting to the American public, the members of the workshop decided. But while the exploration of Mars is a natural and compelling long-range objective, they agreed, trying to do it soon would be unrealistic given resource limitations and more pressing priorities. Instead, humans in space ought to be used to "frame" nearer-term objectives. A couple of members of the panel suggested an objective that is quite a bit closer to home than Mars. They said the civilian space program could make substantial contributions to national and world security by increasing the understanding of threats to Earth by its interaction with the rest of the universe, and specifically, from impacts and changes make by the planet's inhabitants themselves.

The participants in the workshop also agreed that NASA needs a portfolio of activities that should work together at the highest level. Using space to protect Earth in the broadest sense—in several dimensions—would constitute such a portfolio. And *together* is the operative word.

But government has shown no inclination to draw protection strategies together. In the closing years of the twentieth century, and into this one, individuals and groups have moved far ahead of their governments in trying to address potential threats to Earth, both from space and homegrown. One plan, which sprouted almost simultaneously in New York and California, envisions continuously updated cultural and business archives on the Moon as a hedge against natural catastrophes and terrorist strikes against the international corporate community.

On July 8, 2003, a small group of concerned and informed citizens, including Dyson, two former astronauts, and two leading planetary scientists, sent an open letter to Congress. It urged that a comprehensive—and relatively inexpensive—plan to find threatening Near Earth Objects be started; that a means of preventing their slamming into Earth be developed; and that a disaster relief contingency operation be made ready so that if, despite our best efforts, the worst happens, suffering, death, and destruction can be limited. "We write to you today as concerned citizens, convinced that the time has come for our nation to address comprehensively the impact threat from asteroids and comets," the letter began. "A growing body of scientific evidence shows that some of these celestial bodies, also known as Near Earth Objects (NEOs), pose a potentially devastating threat of collision with Earth, capable of causing widespread destruction and loss of life. The largest such impacts can not only threaten the survival of our nation, but even that of civilization itself."

Whatever the reaction (or lack of it) in Congress, the letter was taken so seriously by the Space Foundation, a private sector group in Colorado Springs, that its board of directors passed a resolution on October 27, 2003, endorsing the letter from Dyson and the others and calling on the U.S. and other governments to address the problem. The foundation made the unarguable point that access to space gives humanity the means to protect itself from catastrophe for the first time. It also noted that a number of large asteroids have sped past Earth at relatively close range in recent years. Astronomers get upset at what they consider to be alarmist "close-call" reports and take the position that a miss is a miss. The Space Foundation and similar groups would probably not quibble with that. But they use the NEOs' nearness to make the point that it really is dangerous out there.

There are two basic ways to protect civilization from that danger, as well as the others at home, and both should become the heart of NASA's planetary defense portfolio. One would require the creation of multiple and, in some instances, interlocking ways to physically reduce the peril. A great deal of the necessary technology has long been used. Observation satellites that monitor natural and agricultural resources, devastation caused by nature or people, pollution, ocean hazards, and the creation and proliferation of nuclear weapons have been used for decades. Others, not yet created, could be stationed around Earth as sentinels to give long warning time that asteroids or comets are headed this way. The other way to stave off catastrophe is to spread out. That means migrating to both poles, to very large stations, and returning to the Moon.

To that end, it is crucial that planetary defense not be yet another NASA program, a Planetary Defense Office, boxed onto an organizational chart with all the other offices, projects, and programs. That would condemn it to failure while it was still on paper.

Since the nature of planetary defense is theoretical, at least at present, and since the array of dangers is not perceived as immediate, another program set in the traditional mold would quickly turn into a neglected and starving orphan. Instead, the amorphous and undefined entity that is now called the space program should, almost in its entirety, be focused on one overarching goal: defending Earth with a Planetary Protection Program (no doubt to be called P-cubed by the initiated).

Since the nature of planetary defense in its entirety would be multifaceted and would necessarily have to include the use of force—as would be the case, for example, in deflecting, diverting, or destroying asteroids and comets or preempting a ballistic missile attack—the program would have to bring together the civilian and military sectors in a highly coordinated way. This is not to say both would be joined at the hip to become a single organization. But they would have to be intimately connected where there was a clear commonality of interest. It would be the military's function to intercept cosmic visitors long before they posed a real threat to Earth, and to prevent missile and other attacks not only with technological muscle, but with highly advanced reconnaissance satellites and other machines that do remote sensing. The idea would be to uncover the plans and wherewithal for a nuclear attack anywhere in the world before it occurs and disrupt it in the planning stage or intercept the weapons en route to their targets.

Averting a nuclear attack by any nation would be one of the protection program's most important objectives. The specter of a world-ending nuclear Armageddon that haunted the two superpowers during the Cold War has been succeeded by the increasing possibility that the spread of nuclear weapons among third-world nations and megaterrorists could lead to unprecedented carnage and disastrous regional wars. "Twenty years ago, who would predict that Iraq and North Korea would be the biggest problem?" Richard K. Betts, a political scientist at Columbia University, has said. "The next might also be those with the biggest security problems and the fewest allies, or rising powers which want to be taken seriously. Taiwan? South Korea? Japan? Egypt? Syria? Nigeria?" Some of those countries have both the technological know-how and the ingredients to develop nuclear weapons quickly, and so do several European nations and former Soviet republics. "During the past several years I have spent untold hours in conversation with South Asian planners, and it is clear that they have embarked down the same path as the Cold War rivals in all areas of doomsday planning," Bruce G. Blair, president of the Center for Defense Information, has warned. He cited the Indian subcontinent as particularly dangerous. With limited arsenals to use or lose, any real or imagined attack could quickly escalate into a nuclear war.

No machine can read the mind of a nuclear-armed despot or terrorist. So insiders—spies and informers—would continue to have an important place in the defense of the world against weapons of mass destruction. At the same time, a great deal of information has been turned up over the years by machines in the air and in orbit that looked, listened, and sniffed. Certainly that has been the case with both Iraq and North Korea, but also in many other places, including Israel, where U-2s photographed concrete being poured for a plutonium-reprocessing planet at Dimona in the Negev Desert in 1958 and 1959. The proliferation of superweapons has brought a new urgency, and a formidable challenge, to remote sensing. The various satellite systems, including those using infrared cameras and radar for night and all-weather coverage, would have to be increased as well. And some of the old barriers between civilian and military operations have to come down significantly (although the military and the intelligence communities would, correctly, have to keep many of their secrets, especially in a program that is by definition broadly international).

While keeping Earth safe from cosmic intruders and nuclear warfare is obviously of fundamental importance, as are other active defense measures, they would be only one element in the Planetary Protection Program among many. Maximizing the production of food and other resources would be another facet of the program for which space remote sensing is uniquely suited, and so would protecting the environment by spotting pollution with satellites, making the most efficient and least intrusive use of coastal wetlands and other natural systems, and reducing global warming by, for example, orbiting giant solar collectors that would dramatically reduce the use of fossil fuels.

Establishing a colony on the Moon—a lifeboat—that would grow over time until it was a large, independent outpost is of paramount importance. The lunar base would contain the Alliance to Rescue Civilization's continuously updated archive—not a time capsule—holding virtually every important aspect of life on Earth, from historical, artistic, and scientific records, to life specimens in the form of DNA. ARC would be a repository that would back up life on Earth the way disks back up a computer's hard drive. Were something catastrophic to occur on Earth—say a rogue asteroid hit or a strike by a large comet from the Oort Cloud that could not be predicted—the lunar settlement would constitute the base from which a rescue operation could be launched. And in the worst-case scenario, with the planet annihilated, the settlement and huge stations would ensure the survival of the race and its culture.

The concept of a settlement on the Moon has long since gone beyond the domed Minneapolis model in Clarke's *Exploration of Space*. Most of the logistical requirements, technology, and human factors have been well established by engineers and others in the space agency, in academe, and elsewhere. *The Lunar Base Handbook*, to take one good example among many, gets down to medical and psychological requirements and constraints, power supply, shielding requirements, utilization of lunar resources, logistics, operations costs, and a great deal more.

The first real step in the move to space will start with frequent access to Earth orbit and beyond. That, in turn, will require highly efficient and reusable spacecraft to do in orbit what the C-130 transport planes do when they haul construction materials and life-support cargo to the Antarctic. Constant, total access to space is fundamental to the Planetary Protection Program because it is the only way to get the stations built, colonize the Moon, and then move beyond them to spread the species and therefore greatly enhance the chances of survival.

Ray Erikson has made the point that throughout its existence, humanity has shown itself to be fundamentally incapable of devising and implementing very long-range ways to protect itself from grave threats. He has defined humankind's invariable reaction to dangerous situations as "fight or flee." Now it is time for the civilian and military space programs and their international counterparts to go beyond temporary expedients and instead devise ways to protect Earth and its inhabitants into the infinite future. Access to space—technology—is the easy part. Creating a coherent long-term strategy for survival and adhering to it will be more difficult. But the wherewithal to do so exists. What is needed is the will.

# : 5 :

# A BEEHIVE CALLED EARTH

It is recorded that there was a distant time, now remembered by only a few, when there were no cell phones. It was a primitive time. Communication signals were transmitted, line-of-sight or bounced off the atmosphere, between towering, ungainly antennae that sprouted around the planet. The description and movement of weather was told in the same way, so forecasting was slow, difficult, and inaccurate. There was little or no warning of hurricanes, blizzards, or floods, so many died because of the weather. Old people tell of learning about news either by reading it on printed paper or by listening to it on the radio. Later, they watched black-and-white television pictures that were broadcast directly on "airwaves," like radio signals, from studios to wooden furniture receivers with vacuum tubes and small screens. They navigated in their cars with paper maps and by looking for signs on the sides of roads. Sailors found their way with sextants and stars. In those days, all planes that flew long distances had to carry human navigators who also used maps called navigation charts, and who directed pilots to fly from one ground beacon to the next. The charts themselves were created by airplanes that flew many thousands of flights to record details on the ground. Most astonishing, many countries' real sizes and boundaries were not accurately known. And the locations of islands in the Pacific Ocean, including Hawaii, were off by scores of miles. The Hawaiian Islands, one of the creators of the new system would observe, had inadvertently been "misplaced."

Then everything changed. A new system was born that used the thousands of machines that were sent to space beginning in 1957 to revolutionize humanity's interrelationship and its understanding of the planet on which it lived and the universe beyond.

The foregoing could be written by someone who is so dependent on those machines that he or she knows nothing about existence before them except as history. This is not far-fetched. What is not experienced, be it humans setting foot on another world for the first time, or being killed on this one by spacecraft disintegrating during reentry or trying to get to orbit, can come down only as increasingly remote history. But the satellites are there and can be experienced, as a long look at a clear night sky reveals when the occasional point of light moves silently across the motionless stars. There were more than twenty-seven hundred such points of light in all directions and greatly varying distances as the last millennium gave way to this one. Many were exhausted derelicts, now turned to debris, while many others were working at a variety of tasks for the race that had created them.

Altitude has always brought advantage. Daedalus, the supremely imaginative master craftsman of Greek myth, understood the gift of height. The story of how he escaped from Crete by making giant bird's wings and using them to fly to Sicily has been told since Ovid's time. Medieval architects designed castles with high parapets so boiling oil and other weapons could be dropped on enemy soldiers. A combination of bird envy and canniness sent Chinese soldiers high into the air on kites to spot enemy armies centuries ago. The urge to fly, an inexorable combination of sheer adventure and utility, got balloons off the ground so their occupants could enjoy splendid vistas and military advantage. Inevitably, the longing for more speed and altitude (and the frustration that came with dependence on hot air and the whim of the wind) led to the creation of airplanes that flew ever faster and higher. Then came the rockets that would finally reach space because they didn't have to breathe air to function.

The rocket's potential to get humanity off Earth and to other worlds was perfectly understood by its pioneering theorists, foremost among them Tsiolkovsky. And its unique ability to fling all sorts of machines to extreme altitudes to perform many important but until then impossible missions was clearly envisioned by both fiction writers and engineers between the world wars. Early in 1946—eleven years before the dawn of the space age—scientists and engineers at the Douglas Aircraft Company in El Segundo, California, were assigned to a research and development project that was supposed to evaluate the long-term potential for what were then called artificial Earth satellites. The group, which would soon become the Rand Corporation, completed its study on May 2. The 326-page typed report was called the

*Preliminary Design of an Experimental World-Circling Spaceship* and was one of the seminal documents of the space age. A short introduction predicted that satellites circling Earth would become "one of the most important scientific tools of the Twentieth Century." The authors designed a rocket propulsion system that would get a five-hundred-pound satellite into orbit and noted that one obvious mission for such a machine would be reconnaissance; that is, collecting intelligence from an unexcelled vantage point. The Cold War was already under way.

The Russians, who did pioneering work in rocketry, were every bit as aware of artificial Earth satellites' potential, for both war and peace, as were their new enemies. Beginning in the late 1940s, a group of engineers under Mikhail K. Tikhonravov, the genius who did important early work on multiple stages, studied practical applications for satellites at NII-4, the military rocket research institute. Both Tikhonravov and Sergei Korolyev, his boss, understood the scientific and military value of satellites and repeatedly appealed to the Kremlin for funding to build them. They were ignored for two reasons. For one thing, rocketry was in its infancy and was known to be so unreliable it drew ridicule. (The Jet Propulsion Laboratory, which was founded in 1943, was not named the Rocket Propulsion Laboratory because rockets were broadly associated with comic-strip characters like Buck Rogers and Flash Gordon, and with the Fourth of July.) Second, national defense had the first priority for funding, and that had to do with traditional weapons—ships, tanks, artillery, planes, and other standard military hardware, not with devices that shot out of sight and disappeared. But the rocketeers quietly pushed on.

The situation changed by 1953, largely because the Soviet ballistic missile program needed powerful, reliable rocket motors. (The terms *motor* and *engine* are often confused: motors use solid propellants, while engines run on liquids.) According to Tikhonravov, the advent of the more powerful motors and engines to power increasingly heavy missiles finally made it possible to plan for the development of satellites. On January 30, 1956, he has written, it was decided to build what he called "a sputnik" in 1957 and 1958. "During 1956, in the USSR Academy of Sciences, a number of conferences were held between scientists who specialized in fields connected with space exploration," Tikhonravov continued. "Each conference was devoted to a single aspect of the problem, such as cosmic radiation, the ionosphere, the magnetic field, etc. But always there were present at each conference the following three fundamental questions: What can an artificial Earth satellite contribute to a given branch of science? What instruments should it carry? And who among the scientists would undertake its development? At the time our knowledge of the physical conditions of the upper atmosphere and space surrounding the Earth was quite inadequate. The subsequent discoveries of such phenomena as the radiation belt surrounding the Earth, the magnetosphere, and other findings only confirm this fact." As all the world found out on October 4, 1957, Korolyev, Tikhonravov, and their colleagues got their *Sputnik* to orbit, which started the space age and sent shudders across the United States. What began as a science mission ended as a huge propaganda triumph that took even Nikita Khrushchev by surprise. And eleven months before *Sputnik* went up, even as it was being conceived, Korolyev's engineers were designing a man-carrying spacecraft and thinking about how to create one that would land on the Moon.

Since the uses of spacecraft orbiting Earth are as ubiquitous as the rocket technology that gets them there, the Soviet General Staff and the engineers who produced the space hardware didn't have to read the Rand study to understand that photographing land and sea from high in the sky would have profound strategic consequences. A reconnaissance satellite program called Zenit was therefore probably approved at the outset, in 1956, and after a series of development problems returned the first pictures in the summer of 1962. That was fully two years after its American counterpart, called a KH-1 (for Keyhole) in a program code-named Corona, dropped its own photographs out of space.

In 1956, that pivotal year for the Soviet program, the Russians published a richly detailed, 361-page handbook on artificial satellites called *Soviet Space Science*. It started with Newtonian laws of motion and progressed through the construction, launching, operation, communication, observation, and many uses of satellites. The last included investigating the atmosphere, Earth's magnetic field, micrometeorites and "cosmic dust," astronomical observation, verifying the theory of relativity, biological research, and military applications. The last would include communication and navigation satellites, and very definitely photographic reconnaissance versions such as Zenit and signals intelligence satellites that intercepted electronic traffic.

Like Zenit, the first U.S. satellite reconnaissance program, Corona, used a complicated camera system to take pictures and then fired a long roll of the exposed film out of orbit in a special capsule that was designed to withstand the heat generated as it reentered the atmosphere. Then a parachute popped open and the capsule floated down under it. Zenit functioned the same way, but with one important difference. Zenit capsules came down on the vast

Russian steppes. Corona capsules came down over a section of the Pacific a few hundred miles northwest of Hawaii, so they had to be snagged in midair by U.S. air force transport planes trailing trapezelike cables and then winched on board. The first successful film recovery was made on August 18, 1960. Although the film's resolution, or clarity, was not as good as film brought back by much lower-flying U-2 aircraft, it covered more Soviet and Eastern European territory than all twenty-three U-2 overflights from 1956 to 1960 combined. And the resolution would get a lot better. If Corona had a frustrating disadvantage, it was that it could take a week or longer to get the photographs taken, dropped, delivered to Hawaii, processed, and sent to Washington for analysis.

But that changed on December 19, 1976, when the first of a new breed of spy satellites was rocketed into the sky carrying an electro-optical imaging system that collected and sent back pictures in near real time; that is, virtually as the television-like picture was taken. The new system was variously known as the KH-11, Kennan (later Crystal), Project 1010, or the 5500 series. Whatever it was called, it improved space-based intelligence collection radically by providing high-resolution imagery—objects five inches or so could be distinguished under ideal conditions from a height of seventy miles—as events occurred.

Space reconnaissance had three essential purposes during the Cold War: to collect intelligence, find and pinpoint targets for attack, and monitor compliance with arms control agreements. While it is generally supposed that collecting vast amounts of information about the enemy from space worked to the advantage of the country operating the system—it did and continues to do so—it also played a more subtle role. It eliminated nasty surprises and therefore fostered stability at a time when two opposing military forces faced each other with vast numbers of nuclear weapons on round-the-clock alert. In doing that, space surveillance played a major role in preventing an accidental nuclear holocaust, and it therefore protected Earth from possible annihilation.

The end of the Cold War and the development of new realities in this century, including the internationalization of terrorism, have largely shifted the focus of space-based intelligence collection and, if anything, increased its importance. And imagery—pictures—are by no means all that is collected. Signals intelligence is the other important way to use technology to gather information. Sigint, as it is called by those who gather and use it, has to do with a variety of ways to collect information electronically. It includes using satellites that eavesdrop on communication, including conversations on cell phones, measure the differing characteristics of radars, intercept engineering telemetry during missile tests, and break into coded video transmissions. (The last was done as early as April 12, 1961, and verified less than an hour into the flight that Yuri Gagarin had indeed become the first man to circle Earth in a spacecraft.) Still other satellites, fitted with twelve-foot-long infrared telescopes, watch for ballistic missile launches—fired either in tests or in anger—from 22,300 miles out, where they are "parked" over the same spot on Earth as they keep pace with its rotation. Knowing that these spacecraft would warn of a Soviet or Chinese ballistic missile attack against the United States almost instantly, setting off a response in kind, is credited with preventing such a first strike and therefore with being yet another important source of international stability. In a perverse way, the reconnaissance satellites also prevented all-out war by targeting both sides with unprecedented precision, keeping each in the other's crosshairs. The term mutual assured destruction, or MAD, was coined by Secretary of Defense Robert S. McNamara in the Kennedy administration, and it was indeed assured.

Now the nature of the problem is fundamentally changing, but the requirement to collect intelligence continues. That's because former CIA director Woolsey's dragon has indeed been replaced by two kinds of poisonous snakes: those who develop weapons of mass destruction, and those who commit terror. The Western intelligence community's worst nightmare is that the former will supply chemical, biological, or nuclear weapons to the latter. The leading candidate in the superweapon proliferation category is North Korea. The "hermit kingdom" has gone to great lengths to conceal its superweapons program from the prying eyes it knows are watching in space. At the same time, for political leverage it also hints darkly at having the capacity to use nuclear weapons and has long-range, conventionally armed ballistic missiles at several places that are also under surveillance from the sky. Not so with terrorist organizations such as Al Qaeda, which are more difficult to spot because they are transnational, fragmented, and mobile. It is the difference between being able to photograph a pile of wood or the termites that live in it.

No single method of collecting intelligence on the scale necessary to check the production and spread of superweapons or activity by terrorists can do the job independently. That includes the various, specialized satellite reconnaissance systems, other technical means of gathering information, and people who spy. Rather, a combination of several sources and methods are ordinarily used to collect intelligence, and the results of each are then combined, or "fused," to form a coherent picture that becomes a written National Intelligence Estimate. The techniques run the gamut from using classic informers on the inside—no machine can read Osama bin Laden's mind—to placing intelligence officers posing as diplomats inside embassies, to telephone taps, to using a variety of mechanical devices, many of them truly inventive. Special electronic sensors have been used along North Korea's borders to sniff krypton-85, a gas that is emitted when spent reactor fuel is reprocessed into weapons-grade plutonium, for example. Other sophisticated sensors were set up by newly cooperative Russians in their embassy in Pyongyang in the 1990s.

Still, satellite reconnaissance has played, and will increasingly play, a fundamentally important role in foiling the proliferation of weapons, unconventional and otherwise. To a much lesser extent, it can be used to locate and monitor terrorists. A plan to capture Osama bin Laden in December 1998 was in part based on satellite imagery showing his whereabouts in the Afghan mountains, and National Security Agency satellites eavesdropped on cell phone conversations among top Al Qaeda leaders until they found out about it and began communicating by other means. The metal birds are better at watching weapons proliferators. A. Q. Khan's nefarious and profitable shipments of nuclear components with weapons potential have been closely observed by satellites. North Korea's nuclear complex at Yongbyon, including a vital reprocessing plant, has been scrutinized for years by National Reconnaissance Office satellites that have sent back highly detailed images for analysis by nuclear weapons specialists. The NRO, which is run by the Department of Defense and the CIA, buys and operates U.S. reconnaissance satellites. It had a monopoly on high-resolution space imaging from the time it was created in 1961 until the mid-1990s when Russia, desperate for income from the West, started to sell excellent satellite imagery on the open market. The pictures were useful for large civil engineering projects. And many were bought by well-off collectors of high-tech kitsch. The Clinton administration responded by declassifying Corona and opening the space surveillance field to the private sector. Companies now operate their own imaging satellites and sell all manner of high-resolution pictures, either off-the-shelf or specially taken. On December 24, 2002, the day after the North Koreans reopened the closed reprocessing plant at Yongbyon, breaking seals and disabling International Atomic Energy Agency surveillance cameras in the process, *The New York Times* ran detailed pictures of the nuclear facility taken by Space Imaging,

Inc., a Colorado firm, two years earlier. DigitalGlobe, a California firm, is a competitor that sells imagery with a resolution of two feet. A month later, the newspaper put a story on page one reporting that "American spy satellites over North Korea have detected what appear to be trucks moving the country's stockpile of 8,000 nuclear fuel rods out of storage." The article was illustrated with commercial imagery that showed the entire Yongbyon nuclear complex, as well as high-resolution close-ups of the facility's plutonium reprocessing plant and the building in which the eight thousand spent fuel rods were stored. Five months later, the *Times* ran another page-one story quoting CIA officials as claiming that yet another North Korean nuclear weapons facility had been found by U.S. satellites, this one at Youngdoktong. And on April 23, 2004, the *Times* ran a DigitalGlobe image of the Ryongchon region, near the Chinese border, following the explosion of a train filled with dynamite that killed more than 160 people, injured 1,300, flattened 1,850 homes, and damaged 6,350 others on more than thirty-four acres. Six days later, the newspaper ran dramatic before and after pictures of the area, also taken by the DigitalGlobe satellite. Other space-based imagery has shown many areas in Iraq during the war, including Baghdad, where buildings, streets, and landmarks are clearly visible.

The pictures would have brought smiles to veteran CIA and military reconnaissance specialists who remembered that access to them during the Cold War required security clearances so stringent they were higher than top secret. Even the mere reference to satellite reconnaissance had been forbidden less than a decade earlier. Yet by the spring of 2003, referring to classified space imagery in the news was commonplace: "Spy satellites show a steady stream of activity around the reprocessing plant," the *Times* reported about North Korea on March 1. A year later, to take only one more example among many, it reported that satellite imagery showed what looked like nuclear equipment being smuggled out of Iraq.

Eight months after that, another old taboo was broken when the *Times* ran a story proclaiming, "Boeing Lags in Building Spy Satellites." While a company spokesman declined to comment because of the classified nature of the work, the *Times* reported that Boeing was running more than a year behind schedule and billions of dollars over cost, forcing the NRO to "shift an estimated $4 billion from other spy programs." The new satellites are to be smaller and cheaper than the half dozen huge ones they are to replace, and there would be many of them, the article explained. Reporting that kind of information had been unthinkable when Lockheed Missiles and Space and TRW were building their bus-sized ancestors a decade earlier. And the commercial imagery is now so good, and there is so much of it, that it can be bought by anyone with a credit card. The intelligence establishment itself regularly supplements its own systems' "take" with commercial satellite imagery, and the use of civilian spacecraft for both routine intelligence collection and potential war-fighting is increasing because it's cheaper than maneuvering their classified counterparts and processing the avalanche of digital data that keeps coming down in near real time; that is, almost immediately after the imagery is collected.

Having access to all that imagery would seem, on the face of it, to be clearly advantageous to the United States and its allies, and to the Russian Federation as well. Yet "global information transparency," as the intelligence community calls the explosion of machines that look and listen everywhere, has a decidedly negative side. If the intelligence establishment can in effect use a credit card to buy excellent commercial imagery, so can tyrants and terrorists. "Fundamentally, global information transparency makes it more difficult to achieve surprise, requires better planning, and simplifies strategies for dealing with the United States," an NRO report noted in 2002. "As a result, the possession of, or access to, satellite services and products is potentially destabilizing when hostile parties have access to such information. More ominously, information that can be derived from commercial satellite imagery products can school adversaries in how to conceal their activities from U.S. reconnaissance satellites and may thus deprive the United States of the information that it currently acquires from its own satellites." In other words, if superweapons proliferators and terrorists can buy detailed satellite imagery of their own weapons facilities and camps, respectively, they have the advantage of being able to see themselves the way their enemy sees them and can then conceal their operations accordingly.

However much the government depends on commercial space imagery, it is decidedly not going out of the reconnaissance satellite business, as Boeing's publicly aired travails indicated. To the contrary, the NRO is developing ever more advanced technology as it tries to stay ahead of threats, political and natural. "The space surveillance systems of the future will have to be agile, reliable, precise, and persistent," the NRO report explained. "They will need to work harmoniously with elements of the Defense Department and Intelligence Community. They also will need to support a variety of new missions, including disaster relief, environmental monitoring, and homeland security activities." That strategy—using reconnaissance satellites to monitor problems unrelated to national security in the traditional sense—is old, established, and important. It shows that, where planetary protection is concerned, civilian-military cooperation can have a synergistic effect and save money by not duplicating operations. As early as 1975, a Civil Applications Committee was established at the U.S. Geological Survey to coordinate the use of classified intelligence imagery with such civilian users as the Departments of Agriculture, Commerce, Energy, Interior, Transportation, the Environmental Protection Agency, NASA, the Federal Emergency Management Agency, and the U.S. Army Corps of Engineers. Classified imagery was used not only for civilian mapping and charting, but also for emergency responses to natural disasters such as hurricanes, earthquakes, floods, forest fires. It also monitored the ecosystem. Close cooperation between the intelligence community and the civilian sector is therefore well established and in place for an important part in the collective planetary defense program.

Dramatic advances in miniaturization mean the NRO can launch an armada of the Boeing-made satellites, if and when the design problems are overcome. This will greatly increase the coverage that is necessary to scour the world for ever more elusive targets. So will the ability to watch the target almost constantly, which is what the NRO means by "persistent" coverage. Intelligence analysts like to be able to study their quarry continuously so as to spot signs of change as they occur. But that is next to impossible when it takes sixteen hours for a target to come around again on a planet that rotates once every twenty-four hours beneath a satellite in a ninety-minute orbit. The time can be shortened by using more than one of the current satellites, of course, but significant gaps in coverage remain. That would change if so many of the small satellites, each using miniaturized components and strung out in a line like pearls on a necklace, could be synchronized to send what would in effect be a continuous picture of a target back home for analysis. One would start imaging a target just as its predecessor stopped imaging it, in effect sending a continuous picture down for observation and analysis.

The most important political change that has occurred in space reconnaissance since the Cold War has been its use in shaping public opinion. Proclaiming that satellite imagery showing exceptionally heavy road traffic between Iraq and Syria just before the coalition invasion in March 2003 seemed to indicate that weapons of mass destruction were hurriedly being evacuated, as the head of the National Imagery and Mapping Agency (soon to become the National Geospatial-Intelligence Agency) did that October 2003, was unheard even a few years earlier. It was done to help justify the budget-busting, unexpectedly long war and occupation to the American public.

That same week, a human rights group issued a report in Washington charging that hundreds of thousands of prisoners were being brutalized in thirty-six forced labor camps hidden in the North Korean countryside. Besides eyewitness accounts of torture and starvation, the report cited satellite photographs as evidence the camps existed. A *New York Times* account noted that the 125-page document was released at a time when the Bush administration was showing heightened concern over the North Korean nuclear weapons program.

The satellite pictures of the North Korean nuclear weapons complex itself had the salutary effect of showing many thousands of readers, including those in the diplomatic community and their governments, in the most vivid way, that the nuclear threat posed by Kim Jong Il was real. That, after all, was the government's reason for admitting to the news media that its space "assets" had gotten the goods on him and his program. But, again, the despot himself and his generals would also have seen the pictures, as his United Nations mission did, and that was not salutary. Knowing that Yongbyon and then Youngdoktong were exposed to cameras carried by both commercial and government spacecraft, and even seeing the facilities on page one of Western newspapers, likely convinced the North Koreans to hide still other weapons factories underground. More ominously, detailed pictures that could be bought with American credit cards or simply pulled off the Internet could also be bought with North Korean and terrorist credit cards, providing precise details of potential Western targets.

The revelations about Pyongyang's nuclear weapons program were particularly interesting because, far from denying they were trying to produce nuclear weapons, the North Koreans responded by boasting that they in fact had them, were reprocessing plutonium to produce more, and that an attack on the production sites by the United States would start a nuclear war.

The month the Yongbyon imagery was made public—December 2002—the North Koreans were caught exporting their other contribution to world instability: Scud ballistic missiles. Spanish warships stopped an unmarked and unflagged North Korean freighter in the Gulf of Aden on the tenth, and Spanish marines who boarded the vessel found Scuds with conventional warheads and liquid propellant for them in containers hidden under thousands of sacks of cement. The ship was turned over to the U.S. navy. Since the four-hundred-mile-range weapons were headed for Yemen, a U.S. ally, the freighter was allowed to go on its way. The Scuds were tracked "all the way out" by U.S. intelligence, according to one administration official. That meant by navy patrol planes and at least one satellite. All ship traffic from North Korean ports is routinely monitored by patrol planes and often from space, not only to track shipments of missiles and other military hardware, but to prevent drug smuggling, as has been done in Australia, Japan, and South Korea.

North Korea is emphasized to provide only one example of how a political entity that threatens the international community can be watched and then, at least in part, thwarted from above. There are many others, including the search for Saddam Hussein's superweapons—which turned up none—and Iran's energetic nuclear program. The invasion of Kuwait, which led to the Persian Gulf war in 1991, is another example. But North Korea has played a formidable role in major weapons proliferation. It got its nuclear weapons starter set from Pakistan in exchange for supplying the Pakistanis with ballistic missiles that could reach their enemies in India. A. Q. Khan, the Pakistani scientist who is known as the "father of the Islamic bomb" and who is a national hero, heads the Dr. A. Q. Khan Research Laboratories. He has profited handsomely by selling compact uranium centrifuges not only to North Korea, but to Iran and Libya. The device was invented by Dr. Gernot Zippe in the 1960s to turn raw uranium into fissile material. At any rate, New Delhi in turn developed its own nuclear weapons and long-range missile program to offset those in the People's Republic of China. The Chinese raced to produce nukes because they felt threatened by the Soviet Union. And as is well known, Joseph Stalin and his successors frantically developed superweapons and euphemistically named "delivery systems" because they feared an all-out attack from the West.

But if the proliferation of weapons of mass destruction has not fundamentally changed in more than half a century, the means of tracking the weapons, identifying their sources, and stopping or limiting their movement has. With the two old superpowers having the only eyes in the sky, and with those eyes and the "product" they delivered held in absolute secrecy, the spread of superweapons could for the most part be treated passively or countered through the diplomatic "back channel." Since the information could not be shared for fear of exposing the way it was collected, in other words, the Americans and the Russians had to content themselves with knowing what was happening without interfering, at least publicly. That is now over.

The steadily increasing number of the imaging satellites themselves and their common technology, for both civilian and military intelligence, has put the whole planet under continuous surveillance. That being the case, it is time for NASA, the NRO, and NOAA—the National Oceanic and Atmospheric

Administration, which operates the weather satellites—to combine into a single entity. That notion would make the intelligence types choke on their martinis. But it comes with two caveats. First, since unique imagery could come in that does have to be withheld from the public and other governments, there would be provision for it to be kept out of the reach of those without the necessary clearances. Second, the merger would concern imagery only, meaning the National Security Agency would not be included. The NSA's armada of signals intelligence satellites would necessarily remain as tightly concealed as they currently are. The Civil Applications Committee, now more than a quarter of a century old, has established a solid precedent for unifying the surveillance systems and protecting sensitive information.

The ubiquitous commercial imaging satellites are used far more often for less dramatic political situations than they are for dynamic international developments on the order of the North Korean nuclear weapons program. Their pictures have become standard fare in all the news media, from morning newspapers to the evening news to the Internet. The invasion of Iraq in March 2003 and its deadly aftermath were imaged continuously from space. The pictures were then superimposed on or collated with standard maps to show readers and viewers exactly where an event took place and what its result was. Pictures of the destruction caused by air strikes against Saddam Hussein's palaces and other government structures, and even of the area where he was captured, were featured in the news media. So were the two bombings of the United Nations headquarters in Baghdad and other counterattacks by pro-Saddam Iraqi nationalists. A story in *The New York Times* reporting the second UN bombing was accompanied by a commercial DigitalGlobe satellite picture. It detailed the area where the attack occurred, including the location of the explosion, the nearby UN compound, and the place where the first bomb had gone off a little more than a month earlier. Six months later, to take another example, the *Times* illustrated a story about a car bomb that killed twenty-seven people in central Baghdad with a commercial satellite picture of the neighborhood that had been taken a year and a half earlier. The site of the explosion was pinpointed and so were the Mount Lebanon Hotel, which was destroyed, and an apartment building, which was heavily damaged. *The Wall Street Journal* ran a richly detailed commercial satellite image of Baghdad under the headline "Pinpoint Warfare" and described how precise digital mapping from space is used to find and destroy targets. An accompanying article explained that commercially available space imagery from Russia, France, and Israel had opened the market and that two U.S. companies were selling pictures taken by their own satellites. The picture of Baghdad had been bought from one of them. Meanwhile, the devastation caused by the earthquake that struck Bam, Iran, was also captured by satellite imagery and shown in the news media. By then, pictures taken from space had become an integral part of the world's news and had made faraway places seem closer and more comprehensible.

And at the opposite extreme from using space mapping and imagery to enlighten the public about wars, natural disasters, and other Sturm und Drang, there was the Iva Crider border dispute. Mrs. Crider made the newspapers in April 2004 when it came to light that a satellite-imaging map overruled nineteenth-century human surveyors who'd used stone markers cut with *RI* on one side and C on the other to put her house just on the Hopkinton, Rhode Island, side of the border with North Stonington, Connecticut. Mrs. Crider, a seventy-nine-year-old former school-bus driver, got caught in a jurisdictional dispute when the satellite map showed that, while her house was indeed in Rhode Island, and her deceased husband and two sons are buried in that state, her garage was in North Stonington. The finding set off a jurisdictional dispute involving other residents of the area who also straddle both states. The situation had the potential to change the state taxes the citizens paid, where they sent their children to school, and even such mundane things as telephone area codes. At one point, North Stonington began sending tax bills to Hopkinton residents, which infuriated Rhode Island politicians. But in the end, state officials on both sides decided to resolve the dispute. Hopkinton's state representative called the ground boundaries "sacred" and said residents of the area should decide for themselves which jurisdiction they wanted to call home. "We're going to go with the bounds as they're on the ground, if they're official and undisturbed," said Bob Baron, the Connecticut Department of Transportation's manager of survey operations. "You kind of have to go with that because that's what the people on the border go by. They can't go by an invisible line. The line as described is not necessarily tangible. The line on the ground you can wrap your arms around," Baron concluded, coming out squarely against satellite mapping. That suited Mrs. Crider just fine. "There's nothing wrong with Connecticut. I've got friends there," she told a reporter. "But it's fifteen miles to the post office."

Another important use of space to protect Earth is for dedicated civilian spacecraft to monitor all sorts of resources and handle other assignments that they are uniquely qualified to do. This concept, too, was so obvious it long predated the space age. Here is what Tsiolkovsky saw in 1911 when he peered into the future: "From the rocket we can see the huge sphere of the planet in one or another phase of the Moon. We can see how the sphere rotates, and how within a few hours it shows all its sides successively . . . and we shall observe various points on the surface of the Earth for several minutes and from different sides very closely. This picture is so majestic, attractive, and infinitely varied that I wish with all my soul that you and I could see it."

It was perhaps fitting, then, that the idea of starting an Earth observation program within NASA came from its Office of Manned Space Flight in the mid-1960s. The manned program thought astronauts taking pictures of Earth was a good way to get extra benefits out of the Gemini and Apollo programs. And in 1964 NASA contracted with universities and other government agencies for studies on the usefulness of remote sensing. The U.S. Geological Survey of the Department of the Interior (Gene Shoemaker's eventual roost) suggested in August 1966 that a small orbiting resources observatory be built, and the following month the Department of the Interior publicly announced it was planning an Earth Resources Observation Satellite program, or EROS. The Department of Agriculture also expressed interest in being able to get comprehensive crop data for its own use. Meanwhile, NASA put a high priority on sensor development of special photographic cameras, vidicon television cameras, and scanners to put on the satellites.

But the program got caught in a political morass. Funding was slow in coming because Congress did not grasp remote sensing's unprecedented potential to monitor Earth for multiple purposes, from increasing food production, to spotting environmental trouble such as forest fires, to supporting civil engineering projects. The Department of Defense, on the other hand, grasped the system's potential only too well. It therefore fought for, and initially gained, control of the civilian satellite observation program. But after intense maneuvering, NASA took over the program in 1967.

Civilian Earth observation was in due course turned over to the Office of Space Science and Applications, which soon got into a series of funding battles with the Bureau of the Budget, Congress, and the White House over the program's cost. Bickering over the budget became an annual ritual in the program, whose spacecraft were by then being called ERTS, for Earth Resources Technology Satellite. In 1970, NASA awarded the General Electric Company, which was the prime contractor for the successful Nimbus weather satellite program, a contract to produce ERTS. The spacecraft were almost ten feet high, five feet wide, and weighed a little more than a ton. A large solar panel on either side made them look like ocean buoys with paddles. The imaging apparatus consisted of a return-beam vidicon system consisting of three television-type cameras, and a multispectral scanner. Ground resolution varied between 30 and 262 feet, depending on the object. Bridges over rivers and dirt roads passing through crop fields were easier to distinguish, for example, than stands of trees or rocks and boulders. While that capability was good for starters, it was in stark contrast to the far-better-sighted NRO intelligence collectors in the ultrasecret Keyhole program. The last of the KH-4Bs in the Corona operation had resolutions on the order of six feet. By 1972, when ERTS became operational, improved KH-8s and KH-9s had the power to resolve objects smaller than two feet. That, as noted, is what the civilian Digital-Globe satellite can resolve now.

There were three reasons for the dramatic difference in resolving power. Since ERTS wasn't imaging relatively small military machines such as radar dishes, tanks, and tactical missiles, it didn't need to be able to spot objects that size and smaller. Furthermore, a high-resolution civilian system would betray the capabilities of the NRO's satellites, or so the intelligence people professed. Finally, the Department of State, which saw ERTS become a public relations gold mine around the world after Western Europe, Asia, Africa, and elsewhere began to use its imagery, didn't want the United States to be accused of using civilian remote-sensing systems for espionage. In reality, the fundamentals of optical telescope design hadn't changed since Isaac Newton, so every telescope designer on the planet, with or without a security clearance, could easily calculate to within a few inches what a large telescope could see from seventy or so miles up. It was the public, not the Kremlin, that was kept in the dark.

The scanner filtered images and assigned colors to their parts based on the wavelengths of the light they reflected. In other words, classes of things on Earth—say, oak trees, sand, cabbage, shale, and seawater—reflect differing amounts of light at different wavelengths. They can therefore be separated and identified by their particular reflective patterns, or "signatures." Vegetation usually reflects more green light than red and is very reflective in the infrared end of the spectrum, which registers heat. The multispectral scanner (MSS), in other words, can identify various kinds of vegetation from 570 miles out by the light they radiate: corn (red); hay grass (light green); sunflowers (orange); alfalfa (purple); and bare soil (brown), to take five examples from among thousands. A signature would usually be established with a "ground truth." That is, a spot that produced a particular signature would be checked out by people on the ground, who would superimpose what the scanner registered on the object. If an imaged area was a stand of Norway pine, for example, the system's computers would be programmed accordingly. Then, every time that particular wavelength signature appeared on imagery, the data bank would identify it as Norway pine. Both the vidicon and the MSS pointed straight down.

The first satellite in the series (and the last to carry an acronym that sounded like a gastrointestinal disorder), ERTS 1, was launched on July 23, 1972, and quickly made history. Between its launch and the end of 1974, as it approached the end of its operational life, the spacecraft transmitted some one hundred thousand pictures of three-quarters of Earth's landmass. They were scrutinized by more than three hundred American and foreign scientists, who quickly understood they had a true bonanza of data. Sixteen days after the satellite went into orbit, for example, its scanner sent down pictures of the Salt Lake City region in green, red, near infrared, and infrared, including a color composite that clearly showed vegetation in the mountainous area to the east and urban and rocky areas to the south and west of the Great Salt Lake (which appeared black). Other imagery was used to check the accuracy of a map made by forestry personnel of an area struck by a forest fire in Fiske Creek, California. The satellite imagery, unlike the map, clearly showed places that had not been burned. It also showed more precise boundaries of the damage. Similarly, an image of the Houston area covering hundreds of square miles not only showed the Gulf coast and inland areas in far greater detail than what appeared on a standard map, but it turned up a new lake north of the city. And it got better. Imagery taken on successive orbits that October allowed analysts to piece together a mosaic of all of Connecticut, Massachusetts, and Rhode Island that showed how those states' land was allocated. The picture clearly showed the difference between land used for light and heavy industry; high- and low-density residential areas; developed rural and urban open spaces; agricultural, woodland, and marshland areas; sand and rock outcroppings; transportation routes; and the edge of Long Island Sound and the Atlantic.

On January 14, 1975, ERTS name was changed to the more mellifluous Landsat. Eight days later, a second observation satellite was sent up. Not only was the new applications orbiter named *Landsat 2*, but the acronym that rhymed with *hurts* was so thoroughly purged that its predecessor was retroactively rechristened *Landsat 1*. It kept working under that name, in conjunction with *Landsat 2*, until it was shut down early in 1978. A third Landsat was launched in March of that year, and it was followed by four more of the spacecraft, which literally redrew the map of the world. *Landsat 7* is faithfully monitoring the changing face of this planet as this is written.

The quantity and vast range of hard information Landsats and their close relatives could produce was and remains nothing short of extraordinary and amounted to yet another revolution. In the areas of agriculture, forestry, and range resources alone, the satellites returned data that discriminated between types of crops, timber, and range vegetation; measured crop and timber acreage by species; determined the quality of the land and its biomass; measured the vegetation's health and stress, literally by taking its temperature; determined soil conditions; and assessed grass and forest-fire damage.

By the summer of 2003, farmers came to depend on satellite and aircraft imagery to monitor their crops' health. Plants under stress absorb less infrared light than healthy plants, so they appear to be "warmer" in infrared imagery and can therefore be specially watered or otherwise helped. Other imagery is routinely used to spot pest infestations, plant diseases, and weeds that require special herbicides. But by that summer, the Environmental Protection Agency was planning an even more subtle monitoring system that would differentiate between tracts of genetically modified fruit and vegetables and those that are not modified. The plan rested on the notion that subtle genetic differences between plants could influence the spectral qualities of the light their leaves reflect. The project was being developed because of concern that a transgenic strain of corn that was modified to produce a natural insecticide could cause the development of insects that are resistant to the poison. Farmers are required by law to plant corn that is not genetically modified on at least 20 percent of their acreage, but many were flouting the rule. The EPA was hoping to spot the lawbreakers from space with the new, supersensitive technology.

Applied to land use and mapping, Landsats made and updated maps, categorized land capability, differentiated between urban and rural areas, helped with regional planning, and mapped land-water boundaries, wetlands, and transportation networks. The maps were used, for example, to plan roads through otherwise inaccessible areas in the Andes that were the cheapest and most efficient to build and were the most cost-effective and efficient to use. And the satellite data came in at a tiny fraction of the cost of sending in scores of surveyors and civil engineers. The satellites also mapped floods and floodplains; reported on snow cover and its boundaries; measured glacial features and sediment patterns; and determined water depth. At sea, they spotted living marine organisms; mapped shoreline changes, shoals, and shallow areas; and kept track of potentially dangerous icebergs to protect shipping.

And they could do still more over the oceans, which are in serious trouble. As one reader of the British science magazine *Nature* complained, "There is a crisis in the world's oceans." The reader noted that pollution, overfishing, the introduction of new species, changes to habitats and the earth's changing climate jeopardize marine ecosystems. Not enough research is being done in these crucial areas. But a great deal could be done, relatively cheaply, from space. All of the oceans are now being so depleted by overfishing that stocks, especially of large predator species such as swordfish, tuna, and sharks, are at an all-time low worldwide, and Atlantic cod are close to extinction. In June 2003 the Pew Oceans Commission's eighteen prominent scientists, environmentalists, and fisheries industry officials called for a drastic overhaul of U.S. fisheries policy because of overfishing.

But it isn't all the fault of the United States. Commercial fishing on both sides of the Atlantic and everywhere else has gone from family businesses in which fathers and sons went out with small nets and lines, to industrial trawlers trailing dredges the size of football fields that literally scrape the bottom clean, harvesting an entire ecosystem, including subspecies such as sponges, as the catch of the day. Jeffrey Sachs, director of the Earth Institute at Columbia University and an expert on sustainable development, has said that the major problems with global sustainability are the "failure of international policies to prevent the abuse of the global commons and to develop better technologies to address global scale challenges." Using Landsats and other imaging satellites to monitor overfished areas and spot illegal fishing—Canada has ordered an end to all cod fishing off Newfoundland and Labrador in order to bring back stocks—could help ease the problem.

Observation satellites' contribution to environmental quality has consisted of monitoring the effects of surface, or strip, mining and reclamation; locating, mapping, and monitoring water pollution; finding polluted air and its effect on regions; determining the effects of natural disasters such as earthquakes, volcanic eruptions, and floods; and monitoring the effects of human activities where defoliation, eutrophication of lakes, and other assaults on the environment are concerned. In October 2003, sensors carried on satellites and airplanes were used experimentally to spot such ocean debris as drifting fishing nets that ensnared seabirds, turtles, seals, and other creatures. And Landsat imagery released in the spring of 2003 showed a stark picture of ten thousand square miles of Amazon rain forest that had been ravaged by the creation of pastureland, soybean plantations, and illegal logging in only one year. The pictures showed a startling 40 percent rise in deforestation compared to a year earlier, when seven thousand square miles were wiped out. Using imagery taken in successive years, Greenpeace, the environmental group, warned that the whole rain forest could be destroyed in eighty years if its insatiable attackers are not slowed. Scientists claim one significant reason for global warming is that a fifth of the rain forest is already gone.

The face of the polar ice caps has certainly been changing, too, and because of growing concern about warming, they have been monitored like patients in the infectious-diseases ward. It is probably fair to say that every major news outlet in the country, electronic and print, has run stories about global warming and the consequent melting of the ice caps at both poles. *Science*, the widely respected journal of the American Association for the Advancement of Science, ran a cover issue called "Polar Science" in August 2002 that warned about the changing climates at both ends of the planet and how they could impact the world. The magazine noted that data on the shrinking ice masses have come from on-the-spot inspections by scientists, submarine sonar, and airplanes. But the most compelling evidence has come from satellite monitoring that revealed a 5 percent decrease in ice between 1978 and 1998. All of the data—subsurface, surface, and space-based—have been pumped into computer models showing that the Arctic Ocean's sea ice could lose more than half of its mid-twentieth century volume by the middle of this century.

Satellite imagery clearly showed the disintegration of the largest ice shelf in the Arctic. Scientists reported in September 2003 that a 150-square-mile region of ice known as the Ward Hunt Ice Shelf had broken off Ward Hunt Island, on the Greenland side of the North Pole, and had itself fractured into a jigsaw pattern of floating chunks of ice. Researchers on the scene said they thought the three-thousand-year-old shelf, some of it a hundred feet thick, came apart because of a century-long warming trend that has accelerated in recent decades. Greenland's ice is also melting dramatically around its entire periphery, as comparative satellite imagery shows. If it continues, and there is every reason to believe it will, there will be higher sea levels and widespread coastal flooding.

For sheer dramatic impact, nothing can touch satellite imagery of the ice breaking up except pictures taken by witnesses at the scene. After looking at the first satellite radar images of the whole Antarctic, taken by a Canadian-American spacecraft, Malcolm W. Browne, a *New York Times* science writer, was moved to report, "The beautiful swirls, tracery and arabesques gracing a collection of radar pictures of Antarctica made public last week would be worthy of exhibition by a gallery of abstract art." But, he went on, the imagery contained clues that did not bode well for low-lying parts of the world, including Bangladesh, the Netherlands, and New York City. The Antarctic ice shelf contains 90 percent of the planet's ice, Browne reported, and if just part of it—the West Antarctic Ice Sheet, for example—breaks off and slides into the ocean, the sea level would suddenly rise by seventeen feet, causing a worldwide catastrophe. The satellite radar pictures were uniquely qualified to help:

"Antarctica has been frequently photographed from satellites and aircraft, but in visible light the continent is a largely featureless expanse of glaring white snow covering a thick layer of ice," Browne explained. "By contrast, radar can penetrate clouds and the snow-covered ice surface to reveal subtle contours and shapes that are otherwise invisible."

Three years later, a floating ice shelf the size of Rhode Island disintegrated along the east coast of the arm-shaped Antarctic Peninsula within a month, which is the blink of an eye by glacial standards. Since the huge mass of ice was floating to begin with, it did not raise the sea level, any more than melting ice cubes raise the water level in a glass. But scientists fretted that the breakup was yet another sign of global warming. *The New York Times* picture editors were so captivated by four images taken by NASA's Terra imaging satellite that successively showed the shelf's disintegration, they got them on page one. (*The Washington Post*'s editors, apparently less impressed, ran the same pictures inside and made them smaller.)

Even as the Larsen B ice shelf (as it was called) was disintegrating into hundreds of icebergs during that February of 2002, a new program was going public that dramatically showed people around the world, firsthand, how their collective environment was taking hits, both by their own hand and nature's. That month, NASA's new Natural Hazard Image Service went online with a dazzling parade of worldwide disasters that were captured in color by a number of imaging satellites, including some that watched the weather for NOAA. In the following months, anyone with access to the Internet could in effect look down from space and watch fires raging across northern South America, southern Iraq, northeastern Europe, Mexico, Central America, the Volga River delta in Russia, northern China, eastern India, the middle of Chile, on Kyushu in Japan, in West Africa, and elsewhere. Other pictures showed the destructive beauty of hurricanes and typhoons as they lashed out in great angry swirls, whipping wind and water, and bore down on their targets in both hemispheres. And there were blizzards in Colorado, floods in Missouri, and winds in southeastern Australia so fierce they stripped topsoil off the parched land

and blew it toward New Zealand. Access to three or more disasters at a time is free to subscribers, and every picture shows where on the globe it is happening, the name of the satellite and imaging system that recorded it, and a short description of the mess:

"The MODIS instrument onboard the Aqua spacecraft captured this true-color image of Typhoon Etau—the 10th typhoon of the season—as it was located 470 kilometers south-southeast of Naha, capital of Okinawa Prefecture. Etau was moving north-northwest at 25 km per hour with sustained winds of 165 km per hour (104 mph) and is expected to reach Okinawa Island very early tomorrow morning." The Aqua satellite took Etau's picture on August 6, 2003, and it was posted the next day.

What a change. On September 21, 1938, before meteorology moved to aircraft, let alone spacecraft, a ferocious, fast-moving hurricane slammed into the New Jersey coast, cut across the eastern half of Long Island, and then headed for New England. Weather forecasting in those days was done by looking at the sky, measuring wind velocity and barometric pressure, and then phoning the data to the next weatherman down the line. But that day, the historian William Manchester has written, they didn't even try to phone one another "until the blow had already fallen, carrying the telephone lines with it." The Great Hurricane of 1938, as it soon came to be called, killed ninety people without warning on Long Island alone and caused an estimated $100 million in damage in 1938 dollars. It effectively cut the island in half. The total number of victims was 682, more than half of them in Rhode Island, and the swath of devastation was horrendous.

When Hurricane Isabel struck the East Coast sixty-five years later, almost to the day, residents of the Outer Banks of North Carolina and the surrounding area had a full five days' warning that it was heading straight for them with winds of up to 130 miles an hour. By the time it actually hit its predicted target, on September 18, 2003, with winds of 105 miles an hour, Isabel had been tracked for twelve days. Its long, swirling tentacles had been penetrated by scientists flying as low as two hundred feet above the ocean surface, and by Air Force WC-130H Hurricane Hunters of the Fifty-third Reconnaissance Squadron, which collected computerized meteorological data from high and low altitude, dropped measuring instruments into its eye, and radioed the information to the National Hurricane Center in Miami by satellite relay. The rugged four-engine transport planes flew out of Keesler Air Force Base in Mississippi, typically on eleven-hour missions. Isabel had also been photographed by Edward Lu, a grizzly astronaut on the *International Space Station*, five days before it struck land, and by NOAA's Aqua satellite as early as nine days before it reached the Outer Banks. Starting on September 13, the satellite data were used to fix Isabel's course and warn those in its path that it was approaching. Some on the coast retreated inland with their most precious possessions. Others, typically, decided to ride it out. The important thing was that the decision was theirs. "I'm from California, and all I can say is that at least hurricanes give you warning," said the maintenance director of a church in Virginia Beach, which was severely flooded by Isabel. It isn't the hurricanes that give the warning, though his point about the unpredictability of quakes is, of course, taken.

Isabel directly or indirectly killed two dozen people, pulled the electrical plug on roughly 2 million others, and left a trail of destruction along its path though North Carolina, Virginia, Maryland, Pennsylvania, and northward before it petered out in Canada. The number of fatalities was relatively low, and the number of structures that were saved because they were boarded up, sandbagged, and otherwise secured was undoubtedly high. And the military benefited from the early warning as well. The navy sent forty warships based at Norfolk to safety at sea, and the air force evacuated F-15 fighters and other aircraft from Langley Air Force Base on the Virginia coast as far inland as Oklahoma. No machine in space could have prevented the destruction, but many of the hurricane's survivors owed their lives to advanced computing and to the robots that watch the weather from hundreds of miles above them. It is vitally important not only to spot the hurricane or typhoon well out to sea, but to be able to predict precisely where and when it will hit land, since uncounted billions of dollars can be saved not only by protecting areas to be struck, but by allowing those not in the path of destruction to stay in business.

In that regard, there is a postscript to the saga of Isabel: not everyone appreciated having five days' warning time. Fox News reported early the following week that one or more residents of Myrtle Beach, South Carolina, wanted to go back to only three days' notice about hurricanes heading their way. The two extra days, they complained, cut excessively deep into the tourist business. It's all a matter of priorities.

The Natural Hazards service is an integral part of NASA's larger Earth Observatory, a Web site that carries information about missions and experiments, and satellite-derived news, features, references, and data about the whole Earth. Those who use it can get information on the atmosphere, the oceans, land, energy, and life with the click of a mouse, and they can access feature stories about natural hazards, squeezing water from rock, winter weather in the North Atlantic, and much more.

The idea behind the Natural Hazards operation, according to David Herring, its founder and director, "is to demonstrate the viability of using space-based remote sensors to gather scientific intelligence about climatic and environmental change on our world." The long-term goal, Herring has explained on the Natural Hazards Web site, is to raise people's awareness of changes on the planet and how they are both affected by them and in part cause them.

The Earth Observatory and its Natural Hazards office are based at the Goddard Space Flight Center at Greenbelt, Maryland, and are funded by NASA. Herring said both operations are so intertwined it is virtually impossible to separate their budgets, which totaled $500,000 in fiscal year 2002–3. It's the hardware that's pricey. The cost of the space agency's Earth Observing System, a fifteen-year program that is supposed to build and launch at least a dozen satellites, is just under $7 billion, he added. EOS is the successor to Mission to Planet Earth, a program that started with the first Landsat, and which was designed to study the complex interactions of land, air, water, and life as they affect climate. But data collected by the science satellites in the Earth Observing System are by no means the exclusive domain of scientists. They are regularly used for such applications as monitoring forest fires, which destroy an average of 4 million acres a year in the United States alone. The data from several imaging satellites are in turn channeled to computer models that update active fire maps several times a day. That helps fire managers on the scene decide where to put their equipment. NASA, the U.S. Forest Service, and the University of Maryland have combined resources, in fact, to create a rapid response system that monitors and maps the thousands of forest and brush fires that start every year. It is a stellar example of a cooperative effort to use space for the protection of Earth.

The Earth Observing System's heavily used Terra and Aqua satellites (which carry several sensors, including the MODIS imaging system used on Typhoon Etau) cost $1 billion apiece. The system's third and final satellite, Aura, was sent into a 438-mile-high polar orbit in July 2004. Its six-year mission is to monitor the atmosphere; the very air we breathe. Like most spacecraft that work similar missions (including those that collect the secret intelligence imagery), they fly near-polar orbits because virtually the entire planet rotates under their cameras and other sensors. But the Natural Hazards people go after any good imagery they can lay their computers on, including from such other space agency orbiters as Landsat and the relatively obscure QuikScat, EO-1, TOPEX/Poseidon, and TRMM (for Tropical Rainfall Measuring Mission). Orbview-2, yet another imagery collector, is a commercial satellite. And good imagery, as noted, is occasionally poached from Geostationary Operational Environmental Satellites, or GOES, as Washington's acronym-addicted bureaucrats call them.

The Natural Hazards office tries to find out what's going on in the world any way it can. Herring explained that two or three times a week, the staff pores over many commercial and operational environmental reports to get a quick grip on what's happening everywhere. "Then we compile an internal report, which we call a 'Playlist,' that summarizes the main details about each event," including a synopsis, its exact location, when a satellite is scheduled to pass overhead, and more. The report is in turn e-mailed to instrument teams throughout NASA. They then either assign their own satellites to cover the situation and, in Herring's words, "either make a pretty picture and send it to us, or they will alert my team where we can go get the data and we will make the pretty picture. So, half the battle is knowing where and when to look, and the other half is being able to quickly skim through the data to see if you acquired good data." Sometimes good data are next to impossible to get, Herring explained. For example, imaging floods is especially troublesome because they are often obscured by the clouds that carry the rain that causes them. The pretty pictures were going to some three thousand people around the world by the summer of 2003, he said, and more than ten times that many were tuned in to Earth Observatory. Herring characterized his audience as consisting of the news media, various nongovernmental organizations such as those of firefighters and foresters, and what he termed the "science attentive" general public.

That public now has a window on one aspect of planetary defense because NASA has given it access to the Earth Observatory and, more specifically, to the daily rundown on the multiple hazards afflicting this planet. Even the Department of Defense has gotten into the act. The public was given a rare look at Department of Defense imagery on August 16, 2003, when pictures of the northeast United States taken the night before the great blackout and during it on August 14 by one of its satellites were supplied to the news media. The spacecraft was in the Defense Meteorological Satellite Program, whose main mission is to check cloud cover so the reconnaissance satellites can collect intelligence imagery. Seen side by side, the two pictures, taken only twenty-four hours apart, showed a dramatic difference in lighting. It is a start.

So was the historic agreement by forty-seven nations that met in Tokyo on April 25, 2004, to share Earth observation data, identify gaps in coverage, and come up with ways to fill them. The agreement was reached at the second Earth Observation Summit as an attempt to improve forecasting abnormal weather, understand climate change, and improve the management of natural resources. The idea was to combine the data collected by dozens of observation satellites and other sensors and disseminate it widely and quickly. The process is not as straightforward as it seems, though, because of international politics and other complications. Japan, for example, refuses to disclose fisheries data that could help Korean and Chinese fishermen who ply their shared ocean increase their catch. And the technological leaders of the new system want the developing nations to help pay for it. But the new organization is nevertheless a historically important step, and one that was born out of improved satellite surveillance capability.

As important as it is to spot the disintegration of polar ice, fires in the Amazon, monsoons that bring massive death and destruction to South Asia and elsewhere, deadly diseases on the move on Earth's surface, and other specific threats, there is also the infinitely greater challenge of taking the measure of Earth as a whole, living entity. That, too, is fare for the ubiquitous robots in orbit. They have come with bewildering acronyms, including GEOS and LAGEOS. But they all were designed so this planet's continuously interacting systems could be studied.

GEOS (not to be confused with NOAA's weather-watching GOES) was the Geodynamic Experimental Ocean Satellite, five of which were launched under various names in the 1960s and '70s by NASA and then by the European Space Agency as part of a long-standing international geodesy program. The program has had many facets over the years, but its essential purpose is to study Earth as a single dynamic entity in mathematical detail. The satellites were developed to find out precisely where on Earth geographical features are. They pinpointed the locations of mountains, valleys, rivers, islands, and other places; learned the exact shape and size of the planet itself; measured how its gravity and magnetism vary from one area to another; and studied the big, dynamic picture on and under its surface. That last assignment included observing phenomena associated with earthquakes, fault motions, regional strain fields, tectonic-plate motion, the motion of the poles, Earth's rotation, and solid earth tides (dry land does in fact rise and fall, ever so slightly, as if it is breathing). None of that could have been done without satellites circling the globe. In this case, the spacecraft was called LAGEOS, a nine-hundred-pound sphere that looked like a huge golf ball because it was covered with 426 laser reflectors (hence the *LA* in front of *GEOS*). By bouncing laser beams off the satellite's reflectors and onto Earth, the regions over which it passed could be measured for tectonic-plate and other movement. Tectonics is the study of the structure of Earth's crust, including mountain chains, rift valleys, faults, and other geological features. Plate tectonics has to do with the dozen huge plates, and many smaller ones, that lie within two hundred or so miles of Earth's surface, and which cause earthquakes and volcanoes when they move.

Paul D. Lowman Jr., an imaginative geophysicist at NASA's Goddard Space Flight Center who has studied Earth's dynamics intensively for decades, has paid tribute to the use of space to learn about tectonics. And by space, he means exploring other planets, as well. He has combined the geologic study of Earth with the study of Venus, for example, and shown in almost poetic terms how both studies are mutually supportive.

"Space exploration has . . . assisted the exploration of Earth and has given humans a 'reality check,' so to speak, on plate tectonic theory. People can now see the entire surface of Earth, ocean basins, and continents from space," Lowman has written. "They can now measure crustal motions with once un- dreamed of accuracy with space geodesy." More important, access to space has positively confirmed the essential mechanics of plate tectonic theory. And it has also given scientists a way to study Earth relative to the other planets for the first time.

It has also delivered unprecedented data on Earth's moving mass and grav- ity field by measuring changes in the distance to within a millionth of a meter between two identical spacecraft orbiting 125 miles apart as they pass in for- mation over its shifting water masses and changing terrain. That is a tenth of the width of a human hair. Mass change—the changing physical relationship between the atmosphere, the oceans, and solid earth—is an important part of the changing climate. Topography, composition, and density of Earth's crust, and the equatorial bulging caused by the planet's rotation, all determine grav- itational force at a given place. (As a reporter for *Science News* put it, cleverly, don't go to the north pole if you're concerned about your weight. Since you would be about twenty kilometers closer to the center of the planet because the equator bulges, you'd weigh about a pound more.) The twin satellites, launched on March 17, 2002, carry the relatively friendly acronym GRACE, for Gravity Recovery and Climate Experiment. Fourteen days of data based on their positions relative to each other provided a team of NASA and

German Aerospace Center scientists with the means to make monthly gravity-field maps of Earth that are from ten to a thousand times more accurate than those produced with the help of less accurate satellites during the previous three decades. "The results are applicable to studies of the general ocean circulation and ocean-atmosphere heat and mass exchange," Professor Byron D. Tapley, an authority on satellite orbit studies and a leader of the GRACE team, explained at a meeting in Denver in 2003. "Measurements of continental aquifer mass change, polar ice mass change, and ocean bottom currents are examples of a completely new remote sensing capability whereby we can use satellite measurements to look into the Earth's interior."

And measuring tiny fluctuations in Earth's magnetic field from space might predict earthquakes. Some scientists think the compression of crystalline rocks or the movement of water in fault zones produces very low frequency magnetic fluctuations that could be picked up by orbiting satellites with ultra-sensitive instruments. NASA and the air force think the idea is promising enough to have anted up $1 million to provide ground instrumentation and data analysis to support readings made by a privately owned satellite, appropriately named Quakesat, that was launched on June 30, 2003. And CNES, the French national space agency, thinks the idea has enough potential to warrant the building of a more ambitious (and expensive) satellite. Many seismologists and other scientists have expressed reservations about the notion of satellite-based earthquake prediction because the changes in the magnetic field before a quake are minuscule on the ground and would therefore be extremely difficult to spot from space. But those who are working on Quakesat think its sensors will be able to do the job anyway. If they're right, untold numbers of lives could be saved, especially in countries in southern Europe, the Middle East, and South Asia, where many cities and towns have relatively primitive brick-and-mortar or adobe buildings with wood frames that easily collapse when shaken.

Within three years of NASA's birth, it was given the least practical and most expensive space assignment imaginable: landing astronauts on the Moon. That put the agency's other mandates—conducting science and applying space technology for the benefit of humanity—in a distant second place for years. During the first decade of its existence, science and applications were joined under one roof: the Office of Space Science and Applications, commonly called OSSA. Both came up on the short end for funding, institutional support, and public attention until Apollo's goal had been realized. But once astronauts had repeatedly landed on the Moon, the agency began to concentrate again on the less spectacular, less expensive, but lastingly important contributions access to space could provide. Congress made it clear that it favored applications—practical stuff—over more esoteric science, including "exploration for exploration's sake." In other words, the politicians, a notoriously practical group, understood that their constituents wanted direct benefit from space for their tax dollars a lot more than they wanted data on Earth's magnetosphere, lunar impact craters, Saturn's rings, or solar flares.

In 1969, the year the first two crews landed on the Moon, President Nixon's Space Task Group reflected public sentiment by coming out solidly for applications. "We have found increasing interest in the exploitation of our demonstrated space expertise and technology for the direct benefit of mankind in such areas as earth resources, communications, navigation, national security, science and technology, and international participation," the group's report said. "We have concluded that the space program for the future must include increased emphasis upon space applications." Nine years later, President Jimmy Carter held to the same course, saying he hoped the 1980s would "reflect a balanced strategy of applications, science, and technology development." Applications was first on his list and, in his words, "will bring important benefits to our understanding of earth resources, climate, weather, pollution and agriculture."

By the time Carter, a trained engineer and the most scientifically literate president since Jefferson, issued that statement, a small galaxy of applications satellites was in orbit or being planned. Long before the space age began, it was understood that access to high altitude could provide two fundamentally important benefits. One was enhanced long-distance communication. The other was weather tracking and forecasting.

Earth's curved surface limits line-of-sight communication signals between two points, since the signals move in straight lines. That is why radio and television transmitting and receiving antenne have traditionally been mounted on the highest buildings or towers, including on the tops of the Eiffel Tower, the Empire State Building, and the World Trade Center.

Arthur C. Clarke, as usual, was ahead of his time when he wrote this just before Sputnik went up, "It may seem premature, if not ludicrous, to talk about the commercial possibilities of satellites. Yet the aeroplane became of commercial importance within 30 years of its birth, and there are good reasons for thinking that this time scale may be shortened in the case of the satellite, because of its immense value in the field of communications."

Shortened, indeed. Telstar 1, the world's first commercial spacecraft, was built by AT&T's Bell Labs and launched in July 1962. It transmitted voice, radio, and television signals from as high as thirty-five hundred miles, making it far and away the highest broadcast antenna ever built. Thirteen years later, to take another example among hundreds, RCA's Satcom 1 was parked in a geostationary orbit (now also called a Clarke orbit in honor of the space visionary). It was therefore in effect a 22,300-mile-high "structure" that provided communication coverage to Hawaii, Alaska, and the continental United States. The box-shaped spacecraft, which measured roughly four and a half feet across and was powered by solar cells, brought television news, entertainment, and educational programming to millions of people in cities, towns, and backwaters across the regions it served. Similarly, the International Academy of Astronauts has called for the use of satellites to foster education, telemedicine, and remote sensing in Afghanistan, which has suffered through three decades of war and economic strife. For roughly $750,000, the IAA has said, pilot projects can provide two-way video connections between doctors m Kabul and rural health centers. There is one physician for every fifty thousand people in the country.

President Kennedy's charge to send Americans to the Moon is famous. But what was lost to the public in the excitement of that mandate was that he also made the point the same year, 1961, that communication satellites were of immense potential value. Kennedy's Policy Statement on Communications Satellites recognized their fundamental importance and called for the private sector to develop them as part of an international effort. "I invite all nations to participate in the communications satellite system, in the interest of world peace and closer brotherhood among peoples of the world."

Burton I. Edelson, a former NASA associate administrator for space science and applications, has noted that Telstar was so famous, its technical contributions so significant, and its impact on the public so great that for years its name became generic for communication satellites. That year, Congress acted on Kennedy's mandate by authorizing the creation of the International Telecommunications Satellite Organization, generally known as Intelsat, to encourage the commercial use of communication satellites. The organization officially started in 1964 with eleven member nations. In 1987, 138 countries were members of the organization, and it grew even larger when former Soviet republics and others joined after the Cold War ended. Intelsat spawned its own satellites, called Intelsats, which were built by several contractors for the Communications Satellite Corporation, a private company under government control, which was founded in 1963 as Intelsat's operational arm.

Intelsat 1, also called Early Bird, was a cylinder weighing only eighty-five pounds and measuring twenty-eight inches across. It was launched in April 1965. More than two dozen of the famous satellites were built in five blocks, or groups, in the decades that followed, and they grew progressively larger and more capable. Intelsat 6, weighing nearly two tons, went up in 1986. It had thirty-three thousand voice and data channels and four to relay television signals. By 2003, the spacecraft had grown even heftier and had far more capacity. Intelsat 907, which went up in mid-February on the last of the French Ariane 4 boosters, weighed five tons. It was sent into an equatorial orbit to upgrade communication for Africa, Europe, and much of the western hemisphere. With 19 percent more capacity than Intelsat 605, which it replaced, it amply handled corporate broadcasting, including the quickly growing Internet market.

The international system has long since been supplemented by regional telecommunication satellites, such as the Arab Satellite Communications Organization (Arabsat), and the Asian Satellite Telecommunications Company (Asiasat), to take only two, and by national systems as well. The United States, Canada, and the Soviet Union started their own operations in the 1970s and were followed by India, China, Mexico, Brazil, Japan, France, Great Britain, Germany, Hong Kong, Spain, Italy, and other countries. Many regional and national systems are operated for profit by companies specifically set up to handle space communication.

NASA's communication role as a federal agency has been limited from the start to launching the satellites and being reimbursed for doing so by their owners. It also did research and development on improving instruments, designing larger satellites, more sophisticated guidance and control systems, and more powerful rocket boosters to keep pace with the heavier spacecraft. But that does not mean the government is out of the satellite communication business. As is the case in others areas, including weather observation, the Department of Defense has its own fleet of "secure" satellites, collectively known as the Defense Satellite Communication System, or DSCS (pronounced Discus), which keep the military in touch by two-way voice, Teletype, and digital data transmission around the globe. In addition, the navy operates its own satellite network, called Fltsatcom (Fleet Satellite Communications), which connects ships, planes, submarines, and ground stations by ultrahigh-frequency transmission systems. The navy satellites are also plugged into the national command authority, including the president's office, and carry a separate Air Force Satellite Communications system, or Afsatcom in military jargon.

The multiplication of communication satellites has been enormous. There have been many hundreds launched, flying along routes from relatively low altitude, to highly elliptical Molniya orbits (named after a Russian spacecraft) to far-out Clarke. There are currently sixty-six operational Iridium cellular telephone satellites in six polar orbital planes, meaning that eleven of the satellites, strung out, circle Earth at an altitude of 420 nautical miles on each of the six tracks. Globalstar, a cellular telephone and data communication satellite system, has forty-eight operational satellites and spares orbiting at 760 miles in eight orbital planes. That means 114 satellites provide communication links to mobile users all over Earth all the time. That's why Gladys can get through to her parents in California from her dorm room in Georgia, or when she is rock climbing in Alaska, or working at a summer job on an island far offshore. Other satellites handle only radio traffic, or radio and television, or cell phone, radio, and television. Whole constellations of them relay specific messages, or voice, fax, and data transmission, or video conferences. The European Space Agency operates its own communication satellites, and the Russian Federation alone has had 120 Strela 3 communication satellites in 870-mile-high orbits at one time. Those banks of long-distance telephone operators wearing earphones and plugging their wires into hundreds of holes in switchboards have gone the way of people who stitched fabric onto airplane wings, thanks to the electronic relay satellites that are always overhead. (Fiber-optic cables and microwave landlines are in widespread use for domestic calls, as well, so satellites are far from carrying the whole communication load. Yet they have long since taken over the brunt of it.)

The effect of mass communication through space on the relations of the world is impossible to overestimate. While the term *revolution* is ordinarily overworked, it is fair to apply it here. News, entertainment, and educational programs have, to be sure, been spread through the use of the far-flung spacecraft. They have carried the outside world to isolated hamlets from the Amazon to the Ganges. Their effect as vehicles for spreading information and education can be quantified. What cannot be quantified, however, is the immense role they have played in subtly but continuously showing people around the world that, whatever the effects of rabid nationalism, tribalism, and fanatical religious zealotry, they are united in their shared humanity and innate right to freedom. That is why despotic regimes in Afghanistan, China, Iran, North Korea, and elsewhere have forbidden the use of satellite dishes to pick up foreign broadcasting. And it is why the dishes sprout on rooftops in Tehran, Shanghai, and elsewhere, no matter what the censors dictate.

Global communication would be crucial in the event of a worldwide, or even regional, catastrophe caused by people or nature. That is yet another reason to keep the spacecraft in orbit. And the underlying international unity that comes from their shared common purpose could be a model for the protection of the planet as a whole.

The dark side is that the communication satellites and their surveillance cousins could be put out of operation or severely impaired by the very terrorists they are supposed to help find. It has long been known that a "space Pearl Harbor," as Donald H. Rumsfeld called an attack on American satellites before he became secretary of defense in George W. Bush's presidency, is a serious possibility. A small nuclear weapon carried by a relatively unsophisticated missile like the old Soviet Scud could raise peak radiation levels in low Earth orbit by a factor of three or four. Satellites flying through the region would be pelted by high doses of radiation. The first systems to go would be the spacecraft's attitude-control electronics or its communication link, effectively turning it into a staggering, out-of-control mechanical drunk. Radiation levels could remain high for as long as two years, interrupting the manned spacecraft program and precluding the launch of replacement satellites. Some military satellites, such as reconnaissance types, are hardened to withstand radiation fluxes. Most are not. As nuclear weapons technology and missiles proliferate, the risks of such an attack will increase. Like other kinds of terrorism, a high-altitude nuclear explosion would cause tremendous damage for the investment. It would disrupt the commercial and civil satellites on which much of the global high-tech economy depends. That's the idea.

Meteorology, the other obvious use for satellites, was also understood before the space age began to have direct and obvious benefits to humanity. Weathermen started dreaming about being able to track weather from high altitude and in real time when television and the ability to get cameras to orbital altitude coincided. Not surprisingly, a 270-pound, TV-toting spacecraft named Tiros 1 (for Television Infrared Observation Satellite) was launched in early April 1960. It and a slew of successors were jointly run by NASA and the Department of Defense. That was only two and a half years after Sputnik. During the nearly three months that their batteries lasted, the satellite's two cameras, working at an average altitude of 450 miles, beamed down twenty-two thousand extraordinary (for the time) images. They included masses of clouds sailing across America's springtime sky, many of them bringing rain. It was followed by ten more spacecraft (also called Tiros), four of them carrying infrared sensors. Like communication satellites, the ones that monitored the weather were controlled by a specialized agency—the Weather Bureau, later rechristened the Environmental Sciences Services Administration, or ESSA—with NASA's task again being to launch them and study their operation in order to make improvements. By the time Tiros 10 went up in July 1965 for a two-year mission, it had been decided to start a comprehensive global weather-satellite system run by ESSA and use nine successive satellites bearing that agency's name from 1966 to 1969.

Those early satellites, like the ones in other programs, were steadily replaced by more sophisticated models. The best known was probably a multi-talented bird called Nimbus. Those spacecraft not only watched the weather around the world, they also helped programs in oceanography, agriculture, geology, and hydrology. And they sent back early data on Arctic and Antarctic ice floes, as well. That information, collected in the 1970s, is especially valuable now because it gives scientists the means to compare the relative state of the ice as it is affected by global warming. The Nimbus program also made pioneering, detailed observations of hurricanes "breeding" in the eastern Atlantic before heading toward the Caribbean and the United States. Beginning in 1974, Nimbus was joined by a series of Synchronous Meteorological Satellites. As their name implied, SMS spacecraft were designed to take a wide view of the weather from geostationary orbit. They were in turn replaced by the advanced geostationary GOES spacecraft that belong to the National Oceanic and Atmospheric Administration's National Weather Service, which is the new, high-tech weather bureau.

*Geostationary* is the operative term. In April 1994, a GOES spacecraft was parked over the equator at a spot that gave it a continuous view of the Atlantic from the edge of West Africa to hundreds of miles inside the U.S. east coast. Two years later, a successor was positioned to watch the U.S. west coast and an area of the Pacific that stretched almost to Australia. "With these two satellites, we are literally able to see from the western Atlantic to the eastern Pacific," a delighted Louis Uccellini, the weather service's director of meteorology declared. The satellites and their successors carry imagers that monitor Earth's surface and cloud cover, and a sounder that records atmospheric temperatures and moisture by altitude. (The one over the East Coast sent back a vivid image of a severe winter storm on January 6, 1996, that stretched from northern Florida to New York, and in which particularly cold weather could be seen.)

Demand for ever more satellite-derived weather and environmental information was becoming so great by the turn of the century that yet another series of satellites, this one designed to operate at higher frequencies that would speed data transmission, was being planned. The National Polar-orbiting Operational Environmental Satellite System, known to its handlers as NPOESS (pronounced EN-poe-ess), was being designed as a polar-orbiting spacecraft able to transmit data to anywhere in the world in less than half an hour (it currently takes three hours). "We are finding that our satellites, which started out designed for weather monitoring, are evolving into environmental satellites," explained Captain Craig Nelson, the executive director of the interagency Integrated Program Office, which was designing the spacecraft. This was an evident reference to the satellites' ability to monitor a great deal more than weather. Specifically, they were designed to locate and describe smokestack emissions, acid rain, and other pollutants that attack the environment. "The new groups of users are demanding instantaneous use of data," he added. "We forecast that, by 2020, we will deliver precision data to anyone with a palm-sized communications device."

Space-based weather forecasting is a decidedly mixed bag, even with billions of dollars invested in ground stations and in designing and building all the satellites, launching them, and using advanced computers to try to divine what the endless stream of data they provide means. The system excels at spotting potentially serious sources of trouble like hurricanes, typhoons, and monsoons, in addition to tracking major storm systems like that fierce one in January 1996. In ancient times, there was little or no warning that a hurricane was bearing down on the Caribbean before it landed a devastating left hook on the Gulf Coast and as far north as New England and Canada.

Forecasting the arrival of a killer hurricane three or four days before it reaches land is crucially important. So is tracking severe snowstorms, torrential rains, and long droughts. The satellites and their support structure more than earn their keep by making the big calls with impressive accuracy most of the time. (Late-model GOES types cost $200 million each just to build.) But the regular five-day, or even three-day, forecast by the local meteorologist standing in front of the "satellite map" can be quite another matter. Rather than say, "It is going to rain tomorrow," he or she predicts "a fifty percent chance of rain" or, worse, just "a chance of rain." The alert viewer knows there is also a chance that his or her honor student will become a bank robber; that a nuclear war will start that evening; the mother of all earthquakes will send the West Coast sliding into the Pacific, or a civilization-ending iron asteroid will obliterate the home planet. "A chance of rain" is not very helpful where taking the umbrella out of the closet is concerned. But meteorologists would respond, with justification, by explaining that the ocean-atmospheric system, where weather is generated, is inherently chaotic. Even small errors in the data used to compute the simulations that become forecasts can have rippling effects that grow. The ability to predict a certain weather condition, which they call deterministic, is therefore limited, and the limitation increases with the length of time of the forecast.

Navigation is another service from space on which civilization is heavily dependent. Sailors have been navigating by using the stars at least since the ancient Greeks started plying the Aegean. Their successors on land, sea, and then in the air used magnetic compasses that pointed north to figure out where they were. One problem with the devices was that magnetic north, where they pointed, and true north are not the same place. Another problem was that the farther north their user was, the more the compass's sense of direction became distorted. Space ended that problem as well. Not long after the space age got under way, somebody figured out that radio signals coming from satellites whose location was precisely known could be used to tell people on Earth exactly where they were. With the seven seas as its domain, the U.S. navy had a big navigation problem, and it therefore took the lead in developing a satellite navigation system called Transit in the 1960s. It was followed in the 1970s by the Pentagon's Navstar Global Positioning System, a far more ambitious $10 billion operation that eventually used twenty-four satellites carrying atomic clocks that broadcast the time one thousand times a second. Anyone on Earth with a receiver could use the transmissions from any four of the satellites to pretty much tell exactly where he or she was.

By the middle of the 1990s, the government's $10 billion investment in the Global Positioning System was being refined and leveraged into a $20-billion-a-year industry by a growing legion of people who wanted to know exactly where they were at any time. Former Boy and Girl Scouts who had mastered the art of tying bowlines and sheepshanks, learned the best way to collect rainwater, and carried the old magnetic compasses on their belts now carried $200 GPS receivers bought in department stores. Scientists were using them to track desert tortoise migrations, follow the grazing patterns of sheep, and chart the slow but steady shifting of Earth's crust. And a steadily increasing number of motorists were using GPS receivers built into their cars and trucks to find out where they were. The receivers started as electronic maps on which the vehicle's location was pinpointed, then quickly evolved to voice directions as well: "You have gone half a block too far," the disembodied voice would say with an authority that could be intimidating. "Make a U-turn and go back to Chestnut, then make a right turn." By then, all the satellites were handling assignments that covered the gamut from spotting nuclear explosions and Earth-threatening asteroids, to telling joggers how far they had run, to informing drivers with cell phones and GPS receivers about traffic conditions ahead of them. The system had by then also become indispensable for all the military services and was widely used in the wars in Afghanistan and Iraq.

With some 20 million people using the second version of GPS, which still relied on twenty-four Navstar satellites orbiting 12,500 miles in space, the government in late 2003 began planning a third-generation GPS, or GPS III. The goal was to improve accuracy, reliability, and resistance to jamming, and make the signals strong enough so they would not to be overwhelmed by nearby radio emissions. In addition, the government and U.S. companies were aware that the European Space Agency was planning its own orbital navigation system, called Galileo, and it was unclear whether the two systems would be competitive, compatible, or totally interoperable. Meanwhile, Russia had its own equivalent of GPS on its less well-maintained Glonass navigation satellites. Nextel, Motorola, and other firms were making cell phones with GPS chipsets by the end of 2003, and one enterprising company was even designing what could be the ultimate in human navigation: an implantable GPS sensor that would literally get the system under its user's skin.

Protecting Earth, whether by sampling its atmosphere for dangerous elements, predicting earthquakes, tracing sources of pollution, tracking everything from fires to hurricanes to wandering asteroids and other potentially destructive forces, to monitoring the breakup of ice at both ends of the planet, and drawing its inhabitants closer together through mass communication, is fundamentally important. But they are not the only ways the space-based robots can help their creators. The global warming problem, to take only one example, will require a thorough revamping of the world's energy systems. That means weaning energy users away from the fossil fuels—coal, oil, and natural gas—and substituting clean and efficient alternatives, such as wind power, nuclear reactors, or fusion reactors. The problem with nuclear energy, of course, is that it is a Faustian deal, as the accident at Chernobyl and the long-term spent-fuel storage problem show.

Solar power is an excellent alternative that could be provided by satellites on active, not merely passive, environmental duty. That is, satellites could be used not just to monitor the problem, but to help end it. And they could not only provide that ever-abundant, clean, and powerful source of energy, but they could also work the other side of the problem by deflecting excessive heat from the Sun itself. The obstacle, as usual, is politics—that is, the economics of profit—not technology. The petroleum and related industries, such as automobile manufacturers, have thwarted the large-scale use of solar power at every turn. They have trumped the environment with relatively short-term financial gain.

The idea of using satellites to reflect solar energy to Earth by microwave transmission from relatively low orbit, Clarke orbit, or even from the Moon—generically called space solar power, or SSP—is old. And the means of doing it are generally well understood. More than fifty years ago, Freeman Dyson foresaw the possibility that one day advanced civilizations would place energy-collecting spheres around their suns. They came to be known as Dyson spheres. NASA and the Department of Energy studied a space solar power system in the 1970s that would have used a reflective panel the size of Manhattan to beam energy to a six-mile-by-seven-mile "rectenna" on Earth. Follow-on studies envisioned smaller satellites and more advanced technology. Martin I. Hoffert, a physicist at New York University, and several colleagues have pointed out that energy from the Sun is eight to ten times greater above the atmosphere than it is on Earth's surface because it is undiluted by the planet's atmosphere, rotation, and cloud cover. Furthermore, and quite obviously, the Sun shines all the time in space. Hoffert and the others have made a compelling case for using space solar power (and other alternative energy sources) in place of fossil fuels. If the Department of Energy's projection that worldwide demand for energy will double in twenty years, and then double again in the next twenty, is accurate, SSP could become an important alternative energy source.

In addition, relatively clean helium-3 from the Moon could be used in fusion reactors on Earth, though they come with their own immense technological challenges. And not all strategies require space operations. Superconducting transmission cables could be built worldwide to move surplus electrical power on one continent to places where it is scarce on another.

This is not to say there aren't the inevitable problems, environmental as well as political. These include the effect on birds and airplanes that fly through the solar microwave beams, as well as the effect on telecommunications. Margo R.

Deckard of the Space Frontier Foundation has made the point that the most insidious effect of harnessing solar power could be from building the photovoltaic solar panels themselves, which would create a lot of toxic waste. One way out of the last problem, she has suggested, is to produce the panels on the Moon. Far out.

Hoffert and his colleagues have suggested another way to slow or stop global warming, and it reflects the sort of imaginative thinking for which physicists are well-known: put a twelve-hundred-mile-wide parasol at a point between Earth and the Sun where it would stay put and deflect 2 percent of incident sunlight.

Now (as they say) for the bad news. The beehive analogy is a stretch because no hive could sustain the number of "bees" that circle this planet. Radar and telescopic monitoring of the envelope around Earth by the United States and Russia has turned up about twenty-four thousand man-made objects. Of the nine thousand monitored in space in the year 2000, twenty-seven hundred were taken to be spacecraft, the great majority of them dead. The rest is debris, almost half of which are fragments of launch vehicles that either separated as they put their payloads into orbit, or from accidental explosions. Much of the rest consists of tools, cameras, gloves, explosive bolts, and other remnants of missions. The whole environment, out to Clarke, more closely resembles a permanent, circling hailstorm than bees around a hive. And given the number of satellites that are parked out at Clarke—ones that are part of the vast communication network, and others that watch for a massive ballistic missile launch, or monitor missile and nuclear weapons tests, or eavesdrop on thousands of telephone calls simultaneously, or intercept missile and spacecraft telemetry—Earth has the distinction of having joined Jupiter, Saturn, and the other ringed planets by creating its own multiple "rings" within the cosmological blink of an eye: less than fifty years. This has created a traffic problem, though there is disagreement on how serious it is.

Ray Erikson, the aerospace engineer who has led design, analysis, and materials development projects in both aviation and space programs, thinks it's becoming exceedingly dangerous out there. He has pointed out that space is by definition big, "and the probability of any given spacecraft being hit by a piece of debris large enough to cause *catastrophic* damage is small, but not zero. On the other hand, the probability of being hit by *some* kind of debris on *any* [manned] mission is now 100 percent; hence the immediate (and unfounded) implication of orbital debris in the *Columbia* disaster." And as horrible as it would be if a shuttle orbiter was lost in a collision with junk, Erikson sees a far more profound problem when he looks at the future. "The bottom line is this: If we don't do something to stem the fast-rising tide of orbital debris, we will shroud Earth with a cloud of high-velocity, self-perpetuating shrapnel before the end of *this* century, effectively ending 'The Space Age' for all future generations."

Others are more sanguine. Andrew Turner, an aerospace engineer with Space Systems/Loral in California who has collaborated with Erikson on an important engineering paper having to do with storing data on the Moon, disagrees on the extent of the congestion and the threat it poses: "A self-perpetuating debris cloud is not sustainable at orbits of a few hundred miles, where piloted flights are currently confined, due to drag forces in the upper atmosphere." In other words, a lot of the debris at relatively low altitude is pulled down by gravity, plows into the atmosphere, and burns up. At higher altitudes, all the way out to geosynchronous orbit, retired spacecraft are maneuvered into "graveyard" orbits that are separated by hundreds of miles from the imaginary ring around Earth that is inhabited by working spacecraft. "Cases where two satellites have collided are extremely rare," he adds. Michael L. Fudge, an expert on debris analysis who has studied the threat to communication satellites in great detail, has called the threat "real, but not dire." Fudge, echoing Turner, has noted that current regulations require all communication satellites, commercial and military, as well as other government spacecraft, to be sent down from orbit or parked in the graveyard within twenty-five years after they go dead.

Nicholas L. Johnson tends to agree with Erikson. He is a NASA debris expert based in Houston. Following the maneuvering of the *International Space Station* to avoid a possible collision with an eight-year-old Russian rocket in May 2002, Johnson made the point that an object smaller than four inches could punch a hole in the *ISS*. "If the *ISS* gets hit at a critical spot with one of those, it's going to be a bad day," he said. "A five-centimeter [two-inch] object coming at the space station . . . is going to put a hole in it." As it is, shuttles have often returned with cracked or pocked windshields after being pinged by objects as small as paint chips.

Nor is debris in the usual sense the only danger in near-Earth space. NASA space watchers began seeing a trail of tiny radioactive droplets in the mid-1990s that had leaked out of Soviet radar ocean-reconnaissance satellites. The rorsats, as the spacecraft were called, used nuclear reactors to power the large radar dishes that tracked U.S. naval vessels. The spacecraft were used from

1967 to 1988. But radioactive reactor coolant is thought to have leaked out of sixteen of the thirty-one rorsat reactors that were fired to higher orbits after a mission was completed. As many as 115,000 of the poisonous droplets are now believed to be circling Earth. They represent another orbital debris hazard to spacecraft in low Earth orbit, according to Paula Krisko, another debris expert in Houston.

One likely way to significantly reduce the debris problem in the long term, Erikson thinks, is to limit the number of autonomous satellites that are launched. This is not to say limit the number of satellites. *Autonomous* is the key word. "Approximately seventy-five percent of the equipment on any given satellite has nothing to do with the satellite's mission, it's just for housekeeping. Propulsion, attitude control, thermal control, electrical power and guidance systems are all peripheral items as far as the main payload is concerned," Erikson has noted. Instead, he has proposed something called a skybuss that would supply other satellites with the housekeeping functions they now do on their own. (*Buss*, or *bus*, is the term used for the section of a satellite that holds the housekeeping hardware, including two-way communication equipment, computers, and the other hardware he mentioned.)

"If someone were to provide a platform that supplied all these 'orbital utilities,'" as he calls the housekeeping functions, "then the customer, whether a commercial, scientific or military operator, would only need to launch their own payload and plug it into the 'buss.'" The imaginative engineer envisions as many as fifty skybusses flying in the five most used orbits, acting as mothers that would in turn do for the dependent spacecraft what they could not do for themselves: stabilize them, for example, so they would not need to carry their own stabilization thrusters and hydrazine fuel.

The array of space-based robots attending the home planet would therefore develop into a permanent, in-place force protecting it in many ways, both aggressively and passively. And the machines would function that way in conjunction with a growing human presence in near space as well. The time has come for people and machines to work together to protect Earth.

# : 6 :

# THE ULTIMATE
# FREQUENT-FLIER PROGRAM

Gene Cernan took a last, long look at Earth, then glanced at the barren valley called Taurus-Littrow, which lies at the northeastern edge of the Moon. He knew the boot and tire marks he and Jack Schmitt had left in the dust during three forays in as many days would still be there a million years from that moment. Grubby and tired, but awed by what he had just experienced, Cernan put a foot on the lunar module *Challenger*'s landing pad and grabbed its ladder. Then, ignoring the notes on the checklist tucked in his cuff, he spoke spontaneously to the people of Earth:

"As we leave the Moon and Taurus-Littrow, we leave as we came, and God willing, as we shall return, with peace and hope for all mankind. As I take these last steps from the surface for some time to come. I'd just like to record that America's challenge of today has forged man's destiny of tomorrow." Then he and Schmitt fired their ascent-module engine and roared off the lunar surface and into the dark sky to join Ronald E. Evans, the command-module pilot, for the trip home. It was exactly 5:54:37 eastern standard time on the afternoon of December 14, 1972.

Cernan had no way of knowing that "some time to come" would include the rest of that century and into the next. There was also no way for him to understand that what he and his fellow Apollo 17 astronauts took away from the Moon, besides 257 pounds of rock and soil samples and the memory of an adventure that would last their lifetimes, was a focused space program. By the time they parachuted into the Pacific five days later, the public had long since become disinterested in missions to the Moon. And the space agency that had sent them there was fracturing into a half dozen programs designed not to accomplish a single, clearly articulated, overarching goal for humans in space, but to assure their and NASA's sheer survival. Landing Americans on the Moon, not once but six times, was the greatest and most dramatic feat in the history of human exploration. It was also an overwhelming political triumph. The problem, though, was that Congress and the Nixon White House equated the heart of the space program with sending astronauts to the Moon. That having been accomplished, they saw little or no point in continuing to lavish funds on the space agency the morning after the proverbial night before. Thomas Paine, the space agency administrator, was therefore getting little or no support for the sort of big off-planet engineering projects von Braun and the others had described in *Collier's* in the early fifties. There was a war in Vietnam. Money was tight. And the constituency had shrunk to the point where it was negligible.

The history of the manned space program after Cernan packed up and went home has been one of ongoing, debilitating compromise because of inadequate funding and managerial complacency that has cost lives. *Skylab*, which was put together mostly from Apollo and Saturn parts, was a compromise. The shuttle, which could have been smaller and simpler, was instead an enormously complicated and dangerous compromise between the conflicting needs of various prospective clients. And it satisfied none of them. It was conceived in the Apollo era to be an integral part of the post-Apollo von Braun plan to support a space station that was, in turn, supposed to support humanity's return to the Moon and then the exploration and settlement of Mars. Dr. Brenda Forman, a highly knowledgeable and articulate veteran of the aerospace industry, has made the point that "all this seemed technically possible in the Apollo era: the trouble was that it was politically impossible. It is the central tragedy of Apollo that the political context that made it possible to send humans to the Moon proved utterly indifferent to the vision of their staying there." That is to say, as Forman put it, there was no real constituency for an Apollo follow-on. And that is why, as explained earlier, NASA tried desperately to build a commercial, scientific, and national-security constituency by grossly overestimating the number of flights shuttles would make and underestimating their cost.

Creating a spacecraft that could do a wide variety of missions cost-effectively required a revolution in technology. The shuttle would be the first reusable spacecraft, the first with wings, and the first to have a reusable thermal protection system. It would also be the first to fly with reusable, high-pressure hydrogen-oxygen engines, and the first winged vehicle to transition from orbital speed to a hypersonic glide during reentry. Orbital speed would typically be Mach 25, or 17,500 miles an hour. Reentry was taken to start forty-four hundred nautical miles from the landing site, when the orbiter was at four hundred thousand feet. Its velocity at that point would be about twenty-five thousand feet a second.

The resulting spacecraft was a heavy-duty, immensely complicated combination thoroughbred and workhorse designed to fly a wide variety of missions. Each shuttle was made of more than 2.5 million parts, 230 miles of wiring, 1,060 valves, and 1,440 circuit breakers. "All of it," *The Los Angeles Times* explained in an article titled "Butterfly on a Bullet," "had to function properly at extremes of speed, heat, cold, gravity, and vacuum—the interaction of its parts just at the edge of human understanding and control. From liftoff to landing, the shuttles flew in peril." The *International Space Station*, the other part of the original Space Transportation System, remains a work in progress with no articulated purpose except endless physiological experiments for long-duration missions that remain in the realm of science fiction. (Its proponents say that just being in orbit—establishing a "beachhead" in space—is purpose enough.)

The underlying problem is both institutional and technological. Two shuttle crews have been killed—ironically, one because of intense cold, the other because of intense heat—because of what Richard O. Covey, an astronaut and the cochairman of a panel monitoring the resumption of shuttle flights in the wake of the *Columbia* accident, called "barriers to good decision-making." The panel that investigated the destruction of *Columbia* concluded within four months of the accident that an institutional culture within NASA that plays down problems, and "constraints" from a succession of administrations and from Congress, were the real and pervasive culprits, though they were not as apparent as technological failure. And it called the disintegration of *Columbia* the tip of an iceberg of problems that included communication breakdowns, an increasingly complacent attitude toward safety, constant budgetary pressure, and management shuffling. Pan Am's once heady vision of sending space-liners to the Moon was long gone (and so was Pan Am). Now black humor had it that NASA stood for No Americans in Space at All.

The institutional decay within the space agency became even more apparent a month after the Columbia Accident Investigation Board issued its report. On September 22, all nine members of the Aerospace Safety Advisory Panel, which was established after the Apollo 1 fire killed Grissom, Chaffee, and White in 1967, abruptly resigned. The ASAP, as the group was called, was created specifically to advise NASA on safety. It was criticized in the Senate for failing to anticipate the problems that led to the loss of *Columbia*, and by the accident investigation board as being "independent, but often not very influential." The senators accused the panel of having failed to find warning signs before the *Columbia* accident and said that Sean O'Keefe, the NASA administrator, should reconstitute the panel to appoint recognized safety, management, and engineering experts from industry and academe and provide necessary checks and balances.

In reality, the safety panel was not independent at all, since by federal law its members—including its executive director—were space agency employees. They quit because they were fed up with being trapped in an untenable situation. "If the panel comes out and really upsets NASA, one of its members complained, "it's going to affect his future," putting him "in a bind." The ASAP's chairwoman, Shirley C. McCarty, complained to a newspaper reporter about "a very big sense of frustration." She added that O'Keefe had expressed disappointment with the panel. "One wants to get out of the way when you're in the way," McCarty added. For example, she said, the panel recommended shortly after the *Columbia* accident that crew escape mechanisms be installed in the three remaining orbiters. It also suggested reorganizing NASA so the safety officer for a flight would be more independent. The agency was not receptive to those ideas, she added, and "this set a stage for less effectiveness." It was also symptomatic of the larger problem: a lack of overall focus on one overriding goal instead of a patchwork of projects and programs designed, as long tradition in all bureaucracies had it, to keep the institution intact. Much of it was make-work for astronauts. The Department of Defense and the intelligence community have huge constituencies of citizens who want protection against enemy powers and terrorists. There is no such constituency for space. Focusing on protecting Earth could create one.

What was mentioned only briefly in the accident panel's deliberations, and largely left out of the public discussions about the losses of both *Challenger* and *Columbia*, and the fourteen people they carried to destruction, had to do with the nature of their missions. *Challenger* carried a tracking and data-relay satellite and a booster rocket to get it to higher altitude after it left the orbiter's payload bay. *Columbia* carried scores of scientific experiments, including one to monitor the respiration of an astronaut peddling a bicycle, osteoporosis research, the structure of flame balls, still more experiments on the human body's response to microgravity, and an Israeli experiment to measure atmospheric aerosols over the Mediterranean and the Sahara. And many of the "experiments" had been designed by schoolchildren and their teachers as part of an educational program developed by NASA. One experiment, submitted by the Shoshone-Bannock High School in Fort Hall, Idaho, was called "More Fun with Urine." The idea was to find out whether urine, "space water," as it was euphemistically called by the students and their teacher, could be mixed with paint and dyes to paint a lunar base instead of using precious water. Australian students had an experiment on board designed to see whether spiders would build different webs in space. Children in Liechtenstein sent carpenter bees because they wanted to know whether their ordinarily disagreeable behavior would change in space. Students in a Syracuse, New York, suburb contributed fifteen harvester ants to find out whether their dispositions would change in microgravity. The experiment was called "Ants in Space." One of almost four dozen experiments submitted by New York City children contained scummy Central Park pond water because, the kids were told by their teachers, they might find out whether "magnetic bacteria would get confused by being in space." Still another had to do with seeing whether the magnetic strip on a MetroCard used for public transportation in the city would be affected by exposure to space.

Left unsaid after *Columbia*'s horrifying end, perhaps to spare the astronauts' grieving loved ones more anguish, was that both it and *Challenger* were carrying payloads that could for the most part have gone up on expendable—unmanned—rocket launchers. Faced with being marginalized, and with a restricted budget that demoralized the most dedicated and visionary people in the space agency, NASA stuffed *Columbia* with children's submissions and many adult experiments that were at best marginal to justify its and other shuttle flights and the funding the program brought in. That's what it had done when it had conceived the Teacher in Space program that claimed Christa McAuliffe. And the New Hampshire schoolteacher's and the others' deaths quietly ended an embryonic Journalist in Space program that was supposed to send a newsman (the smart money was on CBS's space-friendly Walter Cronkite) or woman to report the experience of flying through space on a pool basis to the whole world. Mandated to establish a strong American presence in space, but never given the budgetary wherewithal to do so except for Apollo, successive desperate administrators and their aides were forced, and continue to be forced, to make creative public relations—smoke and mirrors—an underpinning of the space program. But public relations, like other kinds of cosmetics, is only skin-deep. So Richard Feynman's admonition after the loss of *Challenger* still resonates: "For a successful technology," the great physicist warned, "reality must take precedence over public relations, for nature cannot be fooled."

The space age is the offspring of the Cold War, and for more than thirty years, the reality was that a politically based technological competition existed between the two great rivals. That competition constituted the essence of both space programs and not only sent humans to another world for the first time, but forced the creation of an array of new machines, many of them mentioned on these pages, that have made the world safer, more productive, and closer-knit in many ways. That process is continuing even as the conflict that caused it recedes into history. But the requirements of the Cold War, now evaporated, have left a void in which decisiveness, boldness, and focus have deteriorated to the point where missions involving human beings going to space have to be justified by endless experiments, including children's, and still more bicycles and equipment to monitor how humans react physiologically to long missions that are not only not in the works, but that are not even on the horizon.

But a new reality transcends the security requirements of the nation-state. This reality, unnoticed by the world's leaders and their vast political and military support systems, is slowly redefining the nature of security and the threat to it. It is a threat to the planet as a whole, not just to its constituent political entities, and it has to be met accordingly. The reality of the threats facing Earth, some of them as old as the solar system, others the work of its inhabitants, make sending humans to space in a focused, integrated, and long-term program imperative. The rationale was explicitly stated by a character named Osepok, the captain of a huge intergalactic space cruiser in Buzz Aldrin and John Barnes's epic novel, *Encounter with Tiber.* "There's not a place in the universe that's safe forever," she tells her crew. "The universe is telling us, 'Spread out or wait around and die.'"

Ray Erikson readily agrees. "With apologies to Thoreau and the Nature Conservancy (of which I am a member), humankind must wage a perpetual war with Nature, chock full of measures and countermeasures, just to survive. Nature thinks nothing of snuffing several thousand people out of existence with earthquakes, tidal waves, and volcanic eruptions annually," he has explained as a prelude to describing the asteroid threat. "Slowly but surely, it is dawning on our collective consciousness that space travel isn't just 'really neat' . . . it isn't *just* very practical. . . . Rather, we have begun to realize that human spaceflight is absolutely essential to our ultimate survival."

The spaceflight Erikson and a number of other realistic scientists, engineers, and managers have in mind is emphatically not an indefinite continuation of what currently passes for a presence in space. That is, although they believe the *International Space Station* could be a valuable step in the right direction, they become impatient at the prospect of endlessly bolting the thing together while eliminating modules and using the remaining shuttles to occasionally ferry water, broiled chicken, and changes of underwear to its three solitary occupants. What is needed, they explain at every opportunity, is a continuous and increasing human presence that builds and inhabits an array of large, self-sustaining space stations and establishes a lunar base that similarly grows into a large, self-contained colony. The stations' and colony's common denominator would be the capacity to support large numbers of people independently so a catastrophic occurrence on Earth, or on one, or even two, of the outposts orbiting it, would not ring down the curtain on civilization. Spreading the seed is the ultimate defense.

That will not be accomplished with the remaining finicky behemoths in the shuttle fleet, which were supposed to have been retired by now, nor with more of them. No one in aerospace seems to believe otherwise. That is why they are scheduled to be forever grounded in 2010. As it is, Major General Michael C. Kostelnik, the space agency's deputy associate administrator for the space shuttle and space station programs—the original Space Transportation System—said after *Columbia* crashed that the remaining shuttles could be flown to orbital rendezvous with the station by two crewmen sitting on ejection seats that would shoot them to safety in the event of trouble. Ironically, Captain Robert L. Crippen and John W. Young sat on such parachute-packed seats when they made that first shuttle flight in *Columbia* twenty-two years before it crashed. But the ejection seats were removed when the orbiters began carrying as many as seven astronauts, four on the flight deck where Crippen and Young had been, and three in a cube deep inside, where ejection seats could not be used. It was considered unfair that those on the flight deck could "punch out" in an emergency, while the others were trapped inside and therefore doomed. The destruction of *Columbia* made it clear that the shuttles had outlived their purpose and had to be replaced by less complicated, more flexible, and simpler spacecraft. One preliminary contender was called the Orbital Space Plane, but it evaporated after President Bush briefly mentioned his vision for space on January 14, 2004.

Erikson has made the point that "it's still *real* hard to get to space. The shuttle needs the power of twenty-five Hoover Dams just to take a minivan load of people and a single tractor trailer's worth of cargo to low orbit. We can manage the power a little more efficiently with more advanced designs, but it's still a *lot* of power." He and many other designers are frustrated because the basis of rocket propulsion, the propellant, has not progressed beyond Tsiolkovsky's chemicals: liquid hydrogen and liquid oxygen.

There was once hope that nuclear energy would propel spacecraft in a project called Orion that was conceived by Freeman Dyson. He imagined a series of precisely controlled nuclear explosions propelling a spacecraft over immense distances, but he did not of course suggest launching it with such an explosion from Earth. In any case, the mere presence of a large nuclear engine of any kind at the Kennedy Space Center would almost undoubtedly draw radical environmentalist protesters, as it did in October 1989, when the *Galileo* probe to Jupiter was being readied for launch on the orbiter *Atlantis*. And it wasn't even an engine that got the protesters worked up. Because Jupiter is so far from the Sun that solar panels would not be able to supply the power needed to work the spacecraft's sensors, two-way communication system, computers, and other electrical equipment, it was provided with a pair of Radioisotope Thermoelectric Generators, or RTGs, which contained heavily shielded plutonium-238. Heat from the decaying plutonium was to be converted to electricity, as had been done on the *Pioneer 10* and *11* missions to Jupiter and Saturn, *Voyager 2*'s sensational Grand Tour of four of the five outer planets, and the Nimbus 3 weather satellite. No matter. The environmentalists picketed outside the space center's main gate, threatened to break in and throw themselves on the launch pad, and loudly warned that if *Atlantis* exploded on launch, as *Challenger* had, the plutonium would contaminate a large part of Florida. And if *Galileo* came in too low and disintegrated on the orbit it would make around Earth before heading for Jupiter, they told news reporters, the whole planet would be sprayed with deadly radioactivity. It was pure balderdash. But it rattled the space agency.

The Orbital Space Plane was just one concept for an advanced spacecraft designed exclusively to replace the shuttle as a people-carrying link between Earth and the *International Space Station*. There have been several others. Until January 14, 2004, when Bush issued his space initiative, the civilian space fraternity saw whatever succeeded the shuttle as a generic spacecraft capable of performing several roles that could, in turn, have led to a larger and faster full-capability supershuttle (or Shuttle Mark II), perhaps to have started flying in 2025. Some engineers conceptualized the supershuttle as a so-called single-stage-to-orbit vehicle that would have taken off vertically or from a runway, or as a two-stage-to-orbit system in which a first stage would have launched vertically, released an upper-stage spacecraft, and returned to Earth. Others conceptualized a hypersonic air-breather that would have taken off from a runway carrying a rocket-powered spacecraft on its back, launched it, and then landed on the runway to be refurbished. The upper-stage rocket, whether on the vertical or horizontal launcher, could have been a modified space plane. And refurbishing the rockets, airliner-style, was and remains broadly accepted as a way to keep both accidents and costs down. Like a malfunctioning airliner, a supershuttle with a mechanical problem could have been flown back down and repaired.

The environment from which the supershuttle might have emerged is a boneyard of once-promising concepts that were abandoned because of a lack of continued support from the government, lack of private financing, or from problems with advanced technology. The delta-shaped National Aerospace Plane, which was called the Orient Express by the Reagan White House, never got off the ground because of technical problems and the inevitable budget crunch. Then the Delta Clipper, a single-stage-to-orbit demonstrator was unveiled, only to take its turn at being killed by the growing preference for two-stage-to-orbit designs and another budget shortfall caused in large part by the space station. Lockheed Martin was then chosen to develop the X-33, which was supposed to fly—though not orbit—to demonstrate reusable-launch-vehicle technology. The RLV failed because of technical problems with its unusually shaped tanks, and a lack of interest on Wall Street in the *VentureStar*, a commercial rocketliner that was supposed to have evolved from the X-33. Meanwhile, a smaller RLV called the X-34 *Pathfinder* ran into major cost overruns because of its new and advanced engine. The X-33 and X-34 were killed in 2001, and so was yet another spacecraft, the X-38, which was supposed to have been an emergency return "lifeboat" for the *International Space Station* that could get seven or eight people back to terra firma in a dire emergency. Ironically, that role went by default to the Soviet Soyuz capsule, which holds only three people.

Enter the X-43A. One hypersonic project did survive, however, and while it was not technically a spacecraft, it could be modified and adapted to become a large reusable craft that could haul large payloads to orbit. It is a hypersonic air-breather and it works. On November 15, 2004, a modified B-52 bomber dropped a winged Pegasus rocket forty thousand feet over the Pacific west of Southern California. The Pegasus in turn carried a pilotless, squat, black, twelve-foot-long "scramjet" called the X-43A to 110,000 feet—a third of the way to space—where it separated and streaked over the ocean at sixty-six hundred miles an hour, or almost ten times the speed of sound, before doing a series of maneuvers and plunging into the ocean. It was designed and built for NASA's Hyper-X research program, which is supposed to develop a jetliner that can fly coast to coast in less than half an hour and anywhere in the world in two or three hours. But the X-43A and its successors are pointing toward orbit.

The idea behind the various plans for an Orbital Space Plane was to develop a relatively small craft that was easier to use than the shuttles, and—of paramount importance—was safer and less expensive. The shuttle program cost roughly $3 billion a year to operate, or a fifth of the space agency's entire budget, with each launch costing roughly $500 million. Furthermore, the orbiters are labor-intensive and difficult to refurbish after each bruising mission. The community would therefore like to replace the elephant with a saddle horse. And that horse might even be able to carry tourists. Space tourism is a charged issue. Some are convinced that hauling tourists like Dennis Tito, who reportedly paid the Russians $20 million for a flight to the space station in a Soyuz in 2001, will open space because of the commercial incentive. Others are dead set against high-rolling dudes paying for trips that don't cover the costs of building, launching, and operating the spacecraft. And still others think the ease of going to space has been greatly oversimplified. They are convinced that serious tourism, as opposed to expensive stunts, is not in the cards for a long time because of daunting technological obstacles.

The whole equation changed—or seemed to change—when Bush announced what was quickly called his space initiative. In brief remarks, he called for returning to the Moon and then going on to worlds beyond, including Mars. He said the three remaining shuttles should be used to complete the *International Space Station* by 2010 and then be retired. They would be replaced, he added, by a Crew Exploration Vehicle that he said would be developed and tested by 2008 and start flying no later than 2014. Invoking the spirit of Lewis and Clark, the president said the new spacecraft would be able to service the *ISS*, and also ferry people to a lunar base. While the aerospace industry, NASA, and the true believers generally welcomed the idea, many objective space experts and several congressmen voiced skepticism because the initiative would have to fly in the face of the largest national debt in American history. "While I'm encouraged by the administration's renewed interest in the space program, their interest doesn't reflect an honest assessment of the fiscal and organizational realities facing NASA and the financial realities facing the country," said Senator Ernest F. Hollings, a South Carolina Democrat. "Disregarding these concerns will further jeopardize the safety of our astronauts, the integrity and viability of our broad American agenda for space and the nation's fiscal health."

Exactly one week after Bush announced his space initiative—on January 21—O'Keefe abruptly killed the OSP to save $6 billion that would be redirected to the new initiative. Until Bush's remarks, Ray Erikson believed the space plane was going to happen and waxed ecstatic about its many capabilities. "Not only will the ability to be launched on either a Lockheed Martin Atlas or Boeing Delta booster . . . provide alternate access to low-inclination orbits, the presence of space launch complexes . . . for both these boosters at Vandenberg will, for the first time, provide astronaut access to *polar* orbits. . . . Combined with the ability to mate OSP with upper stage boosters of various sizes, this provides robust, all-azimuth, all-altitude manned launch capability." Vandenberg is an air force base. Erikson believed, and still does, that relatively inexpensive and versatile spacecraft are the key to the continuous human presence in space that will inevitably lead back to the Moon and beyond.

The lingering question was whether NASA was up to that task or, indeed, should be. Turning a substantial part of the space program over to the private sector has been the subject of serious consideration for years and is a hornet's nest of arguments. Opponents of the idea have noted that the United Space Alliance, a private concern that contracts to operate the shuttles, does not have safety standards to match the government's, which could be why the *Columbia* accident happened. But that is subjective and selective. *Challenger* blew up on NASA's watch.

Five months after the president announced his vision for human space exploration, a special commission chartered by him issued a report that called for implementing it by changing NASA's ponderous bureaucracy to adapt to twenty-first-century requirements, relying more on the private sector, and reestablishing White House oversight of the space program. The nine-member panel was headed by former air force secretary Edward C. (Pete) Aldridge and included such heavyweight thinkers as Dr. Neil deGrasse Tyson.

The sixty-page report, called "A Journey to Inspire, Innovate, and Discover," was largely based on public hearings the commission held across the country. If humans are to return to the Moon by 2020 in preparation for a journey to Mars, the report said, the commercialization of the space program must become the program's primary focus, and the development of a space-based industry will be one of the principal benefits of the journey. The report also said that the space agency has to transform itself from an organizational structure and management style left over from Apollo to meet different requirements: to a "leaner, more focused agency," both at headquarters in Washington and at ten specialized NASA field centers around the country that often have counterproductive turf battles. Finally, the report suggested that a Space Exploration Steering Council be established in the White House to give the president and cabinet direct access and input to the space program and facilitate interagency coordination. Such a body existed under the first President Bush and was headed by Vice President Dan Quayle. It was disbanded by the Clinton administration. Finally, the panel reported that it was uncertain what role the international community should play in the return to the Moon. It did suggest, however, that participation by other countries should be based on the Joint Strike Fighter program. The development of the advanced combat aircraft is a multinational effort in which influence is determined by financial contribution, but with the United States holding the power to make key decisions.

The veneer of adventure and glamour the public associates with space travel is tempered by the insiders' awareness that there are formidable—some would say almost insurmountable—technological, physiological, and psychological challenges to living in space. Fiction writers and space agency publicists have blurred or ignored them in order to sell the dream. That is perfectly understandable. Yet people in government, industry, and academe who work in the space program know the obstacles are truly daunting. David West Reynolds, for one, is far from convinced that the program will follow the script. The Lewis and Clark analogy mentioned by President Bush is absolutely wrong.

More important, likening the settlement of space to the opening of the Western frontier, which has been a cliché since the days of pulp fiction, is grossly simplistic, in Reynolds's view. Settling new territory, developing natural resources and agriculture, and building towns, railroads, and other parts of a civilized infrastructure were what drove the westward expansion. With that accomplished, it was understandable that the infinity of space became the next, and final, frontier. Understandable, Reynolds believes, but wrong because Earth nurtures human life. Space is decidedly hostile to it. He is a close observer of the space program, the director of a San Francisco science-media organization, and the author of a book on Apollo. Reynolds takes strong exception to the high frontier and new West analogies, which he calls "dangerously misleading," and which NASA and the space junkies have peddled from the beginning. He cites former administrator Daniel Goldin as proclaiming, "Send out the Conestoga wagons." We can do no such thing, even metaphorically. That mind-set, Reynolds argues, avoids coming to grips with the fact that space is not only inherently dangerous for life as we know it, but the lack of gravity and air make working there immeasurably more difficult. Living at both poles and in the ocean are better models for living in space than Lewis and Clark, he has written. Whatever the problems the intrepid explorers encountered in their trek across the Northwest, they did not have to wear bulky protective clothing, carry their own food, water, and air, create artificial gravity, and live in heavily insulated structures that would protect them from deadly salvos of ionized radiation. Those who explored the Grand Canyon or panned for gold in California required protection from the weather and predators, human and animal, but not airtight space suits and modules that would hopefully shield them from high-velocity pebbles that could tear through them like bullets without slowing. Space is also an environment barren of virtually everything needed by explorers to sustain themselves. Unlike Lewis and Clark, space travelers cannot easily "live off the land."

Reynolds also believes that making space habitable for large numbers of people in colonies, as proposed by Gerard O'Neill and others, would be so technologically difficult, given the deadly environment, that the expense would be horrendous. He maintains that the cost of mining and collecting other resources in such a hostile environment would far outweigh their value. That, he adds, would keep the much touted private sector away.

"Although it is a technological wonder, the space shuttle completely failed at its primary mission to dramatically reduce the cost of accessing space," Reynolds continues. Similarly, the "materials science" function of the *International Space Station*—perfect crystals, wonder drugs, and the rest—has not attracted private support. He therefore concludes that it will be necessary for the government to lead the move to space. But that, in turn, flies in the face of a public that sees more compelling issues on the ground and a phenomenal budget deficit. Radically less expensive launch systems and smaller, reusable spacecraft would lower the cost substantially, in the view of some famously practical engineers. But David Reynolds's hardheaded approach is a valuable reality check to the magical dreams of the *Star Trek* crowd and the Mars Society.

And going where no man has gone before (or even where he *has* gone before) will come with pernicious physiological and psychological problems, all of which can be overcome by spending time and treasure. The round-trip expedition to Mars and back would take three years and cover 500 million miles. That assumes the intrepid travelers would land and spend time on the surface studying the place. Else, why go?

Among the millions who watched President Bush announce his space initiative were more than a hundred scientists who jammed the Grandstand Sports Bar in Montgomery, Texas. The physiologists, physicists, biochemists, physicians, and others were in Montgomery to work out a way to get Mars explorers back to the home planet alive and well. Some fifty-five threats, from loss of bone mass, to cancer caused by radiation, to depression, have been identified for long-range missions. During astronaut Norm Thagard's 115 days on *Mir* in 1995, he complained about not talking with his family and colleagues enough, that he felt isolated, and that he might not have lasted another three months. A cosmonaut at the fourth Return to the Moon conference in Houston in July 2002 went a step further. Having three or more people in confined quarters for long periods, he said, "is a sure recipe for murder." The tendency for mammals to grow nasty when they live in crowded conditions has been known for many decades. That is why, learning from *Skylab, Salyut,* and *Mir,* the designers of the *International Space Station* provided ample room to work, exercise, relax, and even find seclusion for days at a time. There have been no complaints from its successive occupants on that score.

Beginning in the 1970s, in the wake of Apollo, a low-key debate began over whether the space program should be run by the government or the private sector. It is still going on. Those who think civil servants should steer the program say they are economically unbiased. Those who believe profit-oriented corporations should run it maintain that operating the program for profit would make it more efficient than is possible under the direction of people who collect their paychecks whether the program is healthy or not.

Peter H. Diamandis, a St. Louis entrepreneur and space fan, is squarely in the private sector's camp. In order to jump-start space tourism and get groups and companies into a moribund government monopoly, he announced in 1996 that he would give a $10 million "Ansari X Prize" to the first team that flew a privately financed and built rocket that could carry three people to the fringe of space twice within two weeks. The prize was named for Anousheh Ansari, an Iranian-born multimillionaire who donated more than $1 million to the contest as a way to get others to jump-start the X Prize with $25,000 pledges. More than twenty-five groups and companies decided to compete for the money and prestige, and Diamandis calculated that they would invest about $100 million to win. He also figured that with the concept proven to work, it would then be a matter of gradually extending the altitude and frequency of flights until there was a real presence in space.

Commercial spaceflight literally got off the ground for the first time on June 21, 2004, when a privately built bullet-shaped rocket with stubby wings named *SpaceShipOne* reached an altitude of 328,491 feet to become the first privately developed rocket to reach the edge of space. Space is arbitrarily set at one hundred kilometers, or a little more than sixty-two miles, and *Space-ShipOne* made it with four hundred feet to spare. The rocket was carried to fifty thousand feet over the Mojave Desert by a sleek plane named the *White Knight* and then dropped. Michael W. Melvill, a sixty-three-year-old private test pilot, then took it on a white-knuckle climb to the record altitude. At one point Melvill heard a loud bang. Then *SpaceShipOne* corkscrewed ninety degrees off course in its near-vertical climb and had a mechanical malfunction with the trim system that was supposed to correct for the problem. Melvill therefore had to get back on course manually. After landing, he sat on top of his rocket as it was towed past a crowd of cheering spectators and waved triumphantly at his fans. Later, however, Melvill broke the test pilots' unspoken code by admitting that he had been "deathly afraid" during the close call.

*SpaceShipOne* and the *White Knight* were created by Burt Rutan and his company, Scaled Composites, with $25 million donated by Microsoft billionaire Paul G. Allen. If there is a gene for flying, as some in the fraternity whimsically claim, Rutan has it. As a child in the 1950s, he was captivated by the German rocket designer Wernher von Braun, who appeared on television in Walt Disney's Tomorrowland. "He was on TV with Disney, showing his plan to go to Mars," Rutan recalled on CBS's *60 Minutes* a half century later. "That was life-changing." And like many kids who grew up in that era, Burt Rutan built balsawood flying models. But he was also in a minority that created their own planes. "I challenged myself to design things I didn't know how to design," he recalled. When he grew up, Rutan designed *Voyager*, the graceful, long-winged aircraft that was flown around the world without refueling in 1986, and which now hangs in the National Air and Space Museum in Washington.

Having proven that *SpaceShipOne* could indeed reach space, it captured the X Prize on October 4, 2004, when it touched space for the second time in six days. The first flight was made on September 29 by Melvill, who made it a little higher than sixty-two miles after the rocket unexpectedly spun twenty-nine times as it neared the top of its trajectory. Brian Binnie, who piloted the second flight on October 4, shot to 367,442 feet, or just short of seventy miles, breaking the 354,200-foot record set by test pilot Joe Walker in an X-15 in 1963. Ansari and Diamandis's dream, which had originally drawn snickers from the aerospace industry, had taken off. Other private ventures, two of them bankrolled by dot-com fortunes made in the 1990s bubble, and one by Robert Bigelow, a Las Vegas hotel-chain millionaire, also entered the sweepstakes. Bigelow announced plans at about the time Melvill and Binnie reached space to develop a $50 million prize for the first manned commercial spacecraft that orbits Earth. Binnie and Melvill were both the first to be awarded commercial astronaut wings by the Federal Aviation Administration for reaching an altitude higher than fifty miles. "We've always known that our prize is just a start," said Gregg Maryniak, the X Prize's executive director. An X Prize Cup will be awarded annually, he added.

And then there is Richard Branson, a knighted, eclectic, and flamboyant British multimillionaire who founded the international airline Virgin Atlantic. Now Branson, who thinks really big, wants to take people to space in a new venture called Virgin Galactic. To that end, he was building five spacecraft based on *SpaceShipOne* technology in the Mojave as 2004 came to a close. Branson declared in an interview with *The New York Times* that he would in 2006 take people on three-and-a-half-hour jaunts to space. A round-trip ticket would cost $190,000, he said. And he called his prospective passengers astronauts. "In time," Branson predicted, Virgin Galactic will carry tens of thousands of passengers to space. More than seven thousand have already shown an interest in being taken there, he continued. "We hope to build a hotel in space and offer longer trips, but this will be the start."

Rutan, Branson, and a handful of others are realistic entrepreneurs, not wildly adventurous space junkies. They believe implicitly that they are in the forefront of a truly historic revolution in transportation that is headed inexorably toward carrying people to space for profit. When that takes hold, as it must, the private sector will play an important, if unintentional, role in spreading humans beyond Earth. And both men and their colleagues are openly contemptuous of the aerospace industry and NASA for what they believe is a shameful failure of imagination, courage, and resolve that has settled into a stupor. "We're screwed," one of them happily imagined the Boeing Company and NASA as saying because they know they are being left in the dust and marginalized in the ultimate move off Earth. Yet their small but growing community is necessarily realistic about the hurdles in their way.

A major one, of course, is safety. "Four percent of people sent to space have been killed," Burt Rutan noted. "You can't do that with passengers."

Shuttle and other U.S. piloted missions, including Mercury, Gemini, and Apollo, have been launched eastward from Cape Canaveral (specifically the Kennedy Space Center), a route that has taken them over the Atlantic at low to medium inclinations relative to the equator. There are two advantages to launching in that direction. First, Earth's rotation provides a modest amount of speed for free: the launch vehicle is already moving at nine hundred miles an hour as it sits on the pad, ready to go. Second, the lower stages of expendable launch vehicles and the solid rocket boosters of shuttles fall harmlessly into the Atlantic instead of on populated areas. (The shuttle's external tank, which stays connected to the orbiter longer, is dropped into the Indian Ocean.) Vandenberg, on the other hand, is used for polar or near-polar launches. It is located on the California coastline near Santa Barbara. Lower stages of rockets sent up from Vandenberg fall harmlessly into the Pacific. Polar or near-polar orbits are important for operations that require all, or almost all, of the planet to pass beneath the spacecraft. Space reconnaissance, in which intelligence analysts want potential access to every country on Earth for the collection of imagery, is one such operation. And watching the weather, of course, is another. Neither of those missions requires the use of people in the spacecraft, so operations over the planet's extreme northern and southern regions has been left to the robots. But Vandenberg's use by some kind of space plane, done in conjunction with the Kennedy Space Center, would assure that humans could get to space in any direction. That would provide vastly increased flexibility. But that mission will not be done by the Orbital Space Plane, which, as noted, followed the X-33's and X-34's disappearing act in January 2004, one week after Bush announced his grand, if vague, vision for a "space initiative."

Buzz Aldrin, a realist not given to delusion, has also come out squarely for increased access to space. While many of the men who went to the Moon and those who came after them left the astronaut corps to quietly pursue occupations in other fields, Aldrin remains a steadfast champion of keeping people in space and has stayed actively involved in it for years. His varsity sport at West Point was pole vaulting, the symbolism of which is worthy of Thomas Mann. And he went on to earn a Ph.D. in astronautics at MIT. He also coauthored *Men from Earth*, which told the story of his journey to the Moon, and an autobiography that revealed he couldn't handle being an instant celebrity and suffered a bout of depression and alcoholism. After he recovered, Aldrin coauthored *Encounter with Tiber* and another novel, started a consulting company called Starcraft Enterprises, and then came up with his own Earth-orbiting system, which he named Aquila. Aldrin markets it in a separate company called Starcraft Boosters, Inc.

He started a PowerPoint presentation on Aquila over lunch at a midtown Manhattan restaurant one day in late July 2003 by deploring what he saw as the serious deterioration of the U.S. space program (he called the situation a "crisis"). Then he launched into an unabashedly patriotic call for his country to once again lead the way to space. In quick succession, Aldrin said that the future of the shuttle was unknown (*Columbia* had crashed less than six months earlier and the three surviving shuttles were grounded); NASA was totally dependent on the Soyuzes to reach the *International Space Station;* the station itself was overbudget and years behind its completion schedule; the shuttle's replacements—the X-33 and X-34—had "failed totally"; the Orbital Space Plane (whatever its reincarnation turned out to be) was not to fly until 2012; NASA was losing a major part of its workforce to retirement and has lost more than half of its global market share in commercial launches; and China, with the world's second-largest economy, "will soon launch its first astronaut and has stated its goal to create a base on the Moon." (He was right about the astronauts, as events quickly showed. Less than three months later—on October 15—China became only the third nation to send a person to space when it rocketed a pilot named Yang Liwei into orbit in a Shenzhou 5 spacecraft launched from the Gobi Desert.) All those, Aldrin went on, plus immature reusable-launch-vehicle technology, poorly defined reusable-launch-vehicle requirements, constantly shifting space-station requirements, and the aging bureaucracy and management problems that were reflected in the *Columbia* accident, are symptoms of the *real* problem at the space agency: a lack of direction.

"Shuttle missions basically keep a U.S. presence in space," he continued, but they serve no real exploration goals. "The station is basically only a political tool—a symbol of the end of the Cold War with Russia—that is in the wrong orbit for missions beyond low ones, is too complicated for real space science, and is too expensive for commercialization experiments. "Is it any wonder that the public has lost interest in human spaceflight?" Buzz Aldrin asked rhetorically. "What is interesting about drifting in circles forever?"

Aquila was designed to use the basic space shuttle launch system, including the external tank and solid-rocket boosters, but substitute a variety of specialized pods for the orbiters, making it more flexible. Even the basic pod would be a Delta IV upper stage and would be mounted on the side of the main tank, as the orbiters are. One configuration would hold a crew of eight that could transfer to and from the station, plus up to sixty-seven thousand pounds of cargo. Another would be able to carry ninety-five thousand pounds of supplies and equipment to the station. And still another, smaller version would be able to deliver six people and a three-thousand-pound payload to L1, one of five so-called libration, or Lagrange, points where the gravitational pull of Earth, the Sun, and the Moon cancel each other out, making it ideal for another station because with the three gravity fields canceling each other out, it wouldn't drift. The existence of such regions near Jupiter was first theorized by a French-Italian mathematician and physicist named Joseph-Louis Lagrange, who won a prize awarded by the Paris Academy of Sciences in 1764 for his discovery.

Aldrin said he thought the Orbital Space Plane then envisioned by NASA was not only problematic because of its distant time frame, but was severely limited both in terms of its technology—what it could do—and its growth potential. Aquila, on the other hand, could be put together more quickly because it relied on existing technology. And, Aldrin maintained, it had growth potential, since it could grow into a much larger pod that would be hoisted into space by a pair of Atlas or Delta boosters with wings and tails that would fly back to Earth for refurbishing after the pod was released. Aquila's inherent adaptability would give it the long-term potential to send people back to the Moon "to stay." There, they could rapidly and economically build large space structures such as orbital manufacturing facilities, solar power satellites, and more, according to Aldrin. Some might argue that such heavy dependence on shuttle hardware and infrastructure, at least initially, wouldn't do a lot to advance the idea of an alternate access to orbit. Its creator would no doubt have responded by pointing out that nearly everyone, including General Kostelnik and NASA itself, thought the shuttle as currently constituted had to be replaced as soon as possible. Yet really different approaches to doing so—some would say radical—had been perpetually inert because of traditionally inadequate funding and squabbling over designs within both government and industry. Then Bush's vague space initiative, which called for the development of the Crew Exploration Vehicle to carry astronauts to the Moon and beyond, combined with a manned space program that was in limbo, effectively killed plans for a second-generation shuttle. That Aquila would depend on the extensive use of shuttle technology would seem to make it viable, as opposed to more radical designs that have not been able to get off the drawing board, let alone to orbit. And there have been many of them, all designed to start the human migration to space, and all confined to drawing boards or maybe turned into models adorning executives' desks for want of enough money to pay for the real thing.

NASA, meanwhile, was trying to stimulate radicals, or at least engineers who want to abandon the current shuttle altogether. As Freeman Dyson said, it was trying to stimulate extravagant fantasies that were totally impractical. In April 2003, it awarded $135 million to three competing aerospace groups to work on concepts, technology, and ground-support systems for the Orbital Space Plane, which, nine months later, would go belly-up. The operation belonged in a management textbook as an example of how a governmental agency can squander money. NASA gave $45 million more to the Boeing Company of Seal Beach, California; Northrop Grumman at nearby El Segundo; and a team from Lockheed Martin in Denver; and the Orbital Sciences Corporation, which is based at Dulles, Virginia. They were instructed to come up with other concepts for a space plane. Most of them involved winged craft that looked vaguely like smaller versions of the shuttle orbiter riding to space on top of liquid-propelled carriers and two strap-on solid-rocket boosters. One version, by Boeing, involved the simple, proven space-capsule concept. Dr. John R. Rogacki, NASA's director of space transportation technology, told a *New York Times* reporter in an evident reference to NASA's own thinking about the OSP that the agency was "very flexible in looking for an alternative to get people to and from the space station." He called the designs then under way "technology that is on the shelf or close to it," but made no reference to Aquila. Whatever else it was, the design competition was a playpen for engineers. And it went nowhere.

Andrew Turner has maintained that he has found a way to help solve the problem of frequent access to space by turning the famous credo expressed by Gene Kranz, the mission-control flight director during the Mercury, Gemini, and Apollo programs, on its head. Kranz famously said during the race to the Moon, "Failure is not an option." But Turner thinks "failure *is* an option" where unmanned launch vehicles are concerned and has addressed the perennial cost debacle by giving new meaning to the term *expendable*.

After noting that between 20 percent and a third of electric power is lost during transmission from power plants to consumers, and that roughly the same loss applies to water flowing through aqueducts, he concluded that using relatively cheap launch vehicles to carry propellant or water to space, perhaps losing one-third or more in the process, would be cost-effective. The key is "cheap."

"The cost of lifting a kilogram of mass to orbit is the same regardless of whether it belongs to the revenue-generating telecommunications equipment aboard a commercial satellite, or is merely part of the propellant in the satellite's tanks," Turner explains. "Launch vehicles are designed for high reliability and high payload performance—they are so expensive that they cost almost as much as their cargo." The answer is to send up the propellant, which often amounts to more than half of the mass of the spacecraft, or large quantities of water to orbiting spacefarers, on low-cost rockets that have low reliability and moderate performance. He has named them Aquarius, after the water carrier. "The result," Turner has concluded, "is a lighter, more easily launched spacecraft that receives its fuel after it reaches orbit." In other words, it would in effect be gassed up by a disposable gasoline truck after it reached orbit.

The key is to launch so many Aquarii (as Turner calls more than one of the lifters) that orbiting facilities or depots would be well stocked with consumables. Turner has likened the Aquarius rockets to the vast fleet of low-cost Liberty ships that were used in World War II to carry a stream of supplies to Great Britain and the Soviet Union across an Atlantic that was infested with attacking Nazi U-boats. The mass-produced cargo vessels guaranteed that the total quantity of goods delivered was high and that enough of it would make it to America's besieged allies to sustain them even though material in individual ships was at high risk en route. The transfer of water, food, propellant, or other consumables from the cheap lifters to the spacecraft in orbit would be made by reusable space tugs. Consumable storage could be handled by orbiting depots until delivery was accepted at the space station or by individual spacecraft. While the consumables were being ferried to their destinations, the Aquarius rockets that carried them would be brought down over the Pacific.

Ray Erikson and many of his colleagues see all these as necessary first steps to the stations that will give way to the orbiting colonies and the lunar settlement that will began to spread the seed to avoid extinction. Erikson, Turner, and others are hardheaded engineers, not enraptured dreamers, and they consider themselves realists. Erikson has called the film *Deep Impact* not so much science fiction as "a training film."

"In this context, the *International Space Station* is less a sandbox for scientists than a solidification of the beachhead on the shores of the cosmos established by the Russian space station *Mir* in 1986. In addition to developing the technologies required to defend our planet against impacts, we are learning what it takes to mirror our culture, like a backup disc for a computer, in space, on the Moon, and eventually, elsewhere," Erikson maintains. "With the *ISS*, we have just begun the work needed to establish a permanent, self-sustaining human presence off Earth. This is something we desperately need, not just for fun and profit, not just for a sense of shared adventure, but rather more importantly for our collective continued existence."

However hard creative engineers work at reducing the cost of sending spacecraft to orbit and beyond with chemical rockets—and it will get cheaper—it is still going to be expensive. But there is a much cheaper way, at least in theory: taking an elevator. The concept of using elevator service to get to space was conceived in 1895 by Tsiolkovsky, who envisioned a "tower" thousands of miles high connected to a "celestial castle" orbiting Earth at a constant speed. The centrifugal force of the castle, he calculated, would keep the tower taut the way a rock swinging at the end of a rope keeps the rope taut. The idea appealed to Arthur C. Clarke, who used it in his 1978 novel, *The Fountains of Paradise*. Tsiolkovsky's tower became a cable on which Clarke's spacecraft could ride up to the "castle"—another spacecraft in geosynchronous orbit to which the cable was tethered—and then either be launched into orbit or sent in other directions.

The concept was taken to be theoretically possible, but impossible from a practical standpoint because the strongest element considered for use as the cable, steel, is nowhere near strong enough. Then, in 1991, nanotubes were discovered. These are cylindrical molecules of carbon that are many times stronger than steel. Eight years later, NASA decided to take another look at the space elevator idea, but the concept that evolved was still unworkable. At about that time, Dr. Bradley C. Edwards, who was at the Los Alamos National Laboratory, decided that a nanotube ribbon about three feet wide and thinner than paper attached to a space station could do the job. As he saw it, spacecraft weighing as much as thirteen tons could be pulled up the ribbon by service platforms with a couple of tanklike treads that squeezed the ribbon. The platforms were to be powered by lasers that hit their solar panels. Edwards said at a conference at Los Alamos in September 2003 that all the necessary technology exists to make the space elevator except a long enough nanotube. The longest tube when the conference took place was about three feet. But he said he was confident that a strong enough nanotube-polymer composite would be created in a few years. Edwards estimated the cost of constructing the first elevator at roughly $6 billion, which would be inexpensive considering the nature of the project, not to mention current launch costs. The elevator's advocates, in fact, think it would be dirt cheap, since it would reduce the cost of launching a pound of cargo (or people) from $10,000 to $100.

Erikson believes in the concept and says that the platforms, or "cars," going up the elevator would be balanced by those coming back down, making it, in effect, a conveyor belt. He thinks the cost of getting a pound to geosynchronous orbit would ultimately be counted in pennies. But he also cautions that the nanotube-carbon composite is decades from practical realization and that, in the meantime, cheaper chemical rockets have to be developed. He also worries that a 22,300-mile-long cable would constitute a navigational hazard and be vulnerable to nasty weather.

Erikson, Turner, and their soul mates long ago read Gerard K. O'Neill. While Cold War politics was being played out across the whole spectrum of spacefaring, from Apollo to *Skylab* to the shuttle to space station *Freedom* to the exploration of the solar system, and scientists and managers were arguing among themselves about the relevance of the space program's priorities, others saw the need to go to space from an entirely different perspective. Some, like the anthropologist Margaret Mead and the prolific and endlessly imaginative science fiction writer Isaac Asimov, believed that taking to space fed humankind's need for the sheer adventure of exploration. But others, including scientists from several disciplines who were schooled in the universe's multiple dangers and Earth's finite capacity to sustain the growing number of creatures living on it, took space to be a place where a civilization that existed on a vulnerable and increasingly overtaxed planet could go for safety's sake. The most prominent of them in the 1970s was O'Neill, a brilliant Princeton University physicist, who envisioned huge, populated stations. Carl Sagan, the lyrical astronomer who narrated a wonderfully imaginative thirteen-part television series called *Cosmos*, which came out as a book in 1980, was squarely in O'Neill's corner.

Although Gerard O'Neill specialized in atomic physics and invented the first colliding-beam particle accelerator, he started to develop an interest in humanity's migrating to space and living on colossal stations as early as 1969. That was a year after an international organization called the Club of Rome was started to investigate global change and the prospects for the future of life here. Using relatively sophisticated computer simulation techniques, at least for that time, researchers studied the interaction of accelerating industrialization, increasing pollution, depletion of natural resources, and agricultural productivity. Their controversial report, *The Limits to Growth: A Report for the Club of Rome's Project for the Predicament of Mankind*, painted a gloomy picture of how human activities were adversely impacting the environment, and how activities in one part of the ecosystem have repercussions in the others. The report questioned Earth's ability to sustain life, given steadily growing demands on natural resources that were simultaneously being exhausted and poisoned. In the short term, *The Limits to Growth* predicted, rapid industrialization and increased agricultural output would stay ahead of overpopulation and environmental degradation. But at some point in the twenty-first century, the authors continued, this planet's finite resources would no longer be able to support a large industrial civilization. All of the computer models showed a terrible clash occurring as resources, including food production, and industrialization sharply declined. The study instantly became controversial and made headlines around the world. Some conservatives charged that the Club of Rome was nothing more than a collection of antigrowth, neo-Malthusian doomsayers. The environmentalists, who were relatively new on the scene— the first Earth Day was in 1970—loved it.

O'Neill was no conservative and neither was Sagan. Both agreed that, far from being an impractical fantasy that was not to be taken seriously, moving people off the planet in a continuing migration was the only logical way to reduce pollution, preserve natural resources, and prevent a gradual worldwide collapse that would end with the destruction of Earth itself. O'Neill did not see the imperative of taking to space as romantic, or thrilling, or as satisfying the hackneyed and amorphous (though real) urge to explore. For him, it was emphatically not about having an "adventure." It was about survival. Period.

O'Neill explained by way of rationalizing colonies in space that through the long ages, there were relatively few human beings on the planet, and they had virtually no power over it. Furthermore, whenever a population became too large for the environment to support, it was reduced by war, famine, or plague. That being so, centuries passed without substantial increases in the number of people, and the quality of life for the vast majority was low, even in peacetime. Humanity was so scattered and insignificant relative to the totality of the living Earth, O'Neill theorized, that a creature observing the planet through a telescope from a distant world would have found it difficult to find evidence of human existence.

But all that changed in less than two centuries, or the blink of the cosmic eye, he continued. Humanity suddenly went from being passengers lost on a giant planet to dominating and draining it. Science and medicine, certainly including the rapid development of chemistry, have made fatal diseases rare among infants and children in the better-off nations and reduced their effects in the poorer parts of the world. (That was before AIDS.) "With that one radical change," O'Neill wrote in the front of a book called *The High Frontier: Human Colonies in Space*, which made him well-known almost overnight, "we suddenly find ourselves growing in numbers so fast that Earth cannot long sustain our increase." Meanwhile, he added, our power to change the planet, to force it to accommodate the swelling numbers, has increased to the point where we are profoundly changing the planet and its atmosphere.

That took O'Neill into the industrial revolution and what was, for him, the ultimately fatal paradox he digested in *The Limits to Growth*. The revolution was the mechanism by which a large fraction of people finally achieved a high standard of living. That is, those in the most advanced countries began to live comfortably, had a reasonable life expectancy, were free to travel, and began to have relatively easy access to news, entertainment, and education. But there was a pernicious side to the revolution. O'Neill called it evil. "Though it began only two hundred years ago . . . its side effects have already altered Earth in frightening ways. It has scarred, gutted, and dirtied our planet to a degree that many people find intolerable. Smoke and ash from factories in England cloud the air as far away as Norway, and pollution from the industries of Japan can now be detected in the snows of Alaska." To environmental degradation, he went on, add sharp limits on food, energy, and materials at a time when two-thirds of humanity is still poor, desperate, and on the edge of starvation. What can save them is still more industrialization, including mechanical farming, that will inevitably multiply the attacks on the environment in what has become an increasingly vicious cycle, O'Neill wrote.

"As we strive to find physical solutions to the problems faced by mankind, we must realize, with humility, that we can offer no panaceas. There are no Utopias," the physicist argued. "At the most we can suggest opportunities whose technical imperatives will make it easier for mankind to choose peace rather than war; diversity rather than repression; human simplicity rather than inhuman mechanization. Technology must be our slave, and not the reverse."

The reference to diversity rather than repression alluded to another of the Club of Rome's predictions: the world of the twenty-first century would be so pressed for resources that growth would have to be stopped and the resources rationed. That, in turn, raised the specter of a radical change in American, Western, and even world politics in order to enforce the rationing. In the words of Hans Mark, a friend and fellow physicist, "It was O'Neill's contention that ultimately the only political system capable of performing this function would have to be authoritarian or fascist in nature. If there was no scope for growth and for expansion, then tyranny was the only way to run society."

That notion was reinforced by an influential economist, Robert L. Heilbroner, who had warned about the consequences of what he, too, saw as a fast-approaching severe scarcity of resources and the potentially catastrophic consequences that would be caused by it. In *An Inquiry into the Human Prospect*, which came out in 1974, Heilbroner predicted that what he saw as a time of extreme "goods hunger" would start a large-scale "social reorganization," or a redistribution of the world's output and energy. It was an academician's way of saying that people who would later be called the have-nots would go after the resources of those who had them. And they would use violence if necessary. It would amount to Marx's bloody revolution with many times the ferocity. O'Neill read and absorbed Heilbroner, too. "Under these conditions," he wrote, "Heilbroner feels that the threat of nuclear war is likely to increase greatly in the next decades." He quoted the economist as predicting that "massive human deterioration in the backward areas can be avoided only by a redistribution of the world's output and energies on a scale immensely larger than anything that has hitherto been seriously contemplated." Nuclear "terrorism," he quoted Heilbroner as having warned, "for the first time makes such action possible."

O'Neill maintained that freedom and democracy were viable only if people believed that growth was possible and that new horizons were out there to be reached. Growth for the inhabitants of a planet with burgeoning population and finite resources led Gerard O'Neill to the same conclusion as Tsiolkovsky, and one that was obvious: start colonies elsewhere. Like any number of other bright scientific dreamers, he, too, had read the Russian genius's *Beyond the Planet Earth*. And like them, he correctly characterized the book as a thinly veiled treatise on basic physics. But it captivated him. "It should be read for what it is: a daring but logical feat of the imagination," he said. "At a time when transportation was still almost exclusively horse drawn, it required a bold thinker indeed to speak casually (and accurately) of the necessary orbital speeds of kilometers per second." Like Goddard, Oberth, Korolyev, von Braun, and the other great rocketeers, Konstantin Tsiolkovsky was Gerard O'Neill's patron saint.

O'Neill stimulated the imaginations of the students in his freshman physics class by challenging them to conceptualize a space colony with him. They were apparently fascinated by the exercise, and after assigning it for three or four consecutive years, he published an integrated, overall concept in the September 1974 issue of the prestigious *Physics Today*. The article was so well received it led to his enduring masterwork, *The High Frontier*, which he published himself in 1976. It was commercially published the following year and reissued in a third edition in 2000 by the Space Studies Institute in Princeton to mark the new millennium.

On January 19, 1976, the year he published his book, O'Neill took his dream to Washington in the hope of convincing those who oversaw the space budget that the colonies would help save the United States and the rest of the world. He testified before the Senate Subcommittee on Aerospace Technology and National Needs about the vital long-range importance of using large satellites to direct solar power for use in manufacturing operations in huge colonies in distant orbits. That O'Neill was invited to testify before Congress spoke to the fact that by then, his vision and insight were appreciated, even by politicians. That nothing came of the testimony spoke to the fact that the politicians were unable to process a plan that represented a bold, creative, and long stride into the future. The imaginative physicist ran into three overlapping obstacles that were decisive in causing him to be ignored. Society could not comprehend danger, no matter how menacing, that was theoretical. In addition, the society's collective mentality was utterly incapable of really long-range planning; of preparing for long-range threats. And he had to contend with politicians who understood that activities in space were low priorities for an electorate that struggled with more immediate challenges. There was, in short, no constituency. Colonizing off the planet was therefore an idea whose time had not yet come, politically as well as socially.

O'Neill called his colonies "islands," which was appropriate for the ocean metaphor often used to describe space, and claimed the technology for building them existed when he wrote his book. He envisioned three islands to start with, one relatively small, one colossal. He was also careful to explain that the concepts he imagined were flexible, and that the distinction between science fiction and reality had to be kept in mind constantly. On that score, O'Neill stipulated that only tools then available would be used to construct the islands, not fantastic creations of the imagination, and that the designs were subject to change. "No one will be more surprised than I if, when Island One is completed, it looks very much like the sketches we now make of it." He also insisted that the humanitarian foundation of the program was important. As the physicist saw it, an industrial system that could adequately serve the growing number of people on the larger of the islands for a long time without causing environmental damage was crucial. It would make full use of unlimited solar power and the "virtually unlimited resources of the Moon and the asteroid belt." O'Neill stressed that to be habitable in the best and most important sense of the term, the islands would have to be congenial and comfortable places in which to live, not cold and cramped environments that looked as if they came from the set of *2001: A Space Odyssey.* Some of the illustrations in *The High Frontier* showed people in open shirts and dresses talking near a stream. A nearby town could have been almost anywhere in the United States. Other people were shown in bathing suits at a beach.

O'Neill also had to address dangers that are peculiar to living in space. The most frequently mentioned threat raised by people in the audiences to which he spoke about the colonies had to do with meteoroids. Speeding rocks that could puncture an island's outer wall and perhaps go all the way through it would, after all, cause anxiety. It is understandable that people would wonder about the effect of rocks hitting one of the large, airy space abodes at high speed. Doing the kind of homework the Spaceguard astronomers would do two decades later, O'Neill learned that most of the things speeding around the solar system are cometary fragments, meaning they are dirty ice, or what he called mini-snowballs. And like Morrison, Yeomans, Chapman, and others, O'Neill found that the chances of being hit by a large one are far less than by progressively smaller ones. "Averaging the data from what seem to be the most reliable sources," the physicist reported, "one finds that in order to be struck by a meteoroid of really large size, one ton, a large Island Three community [the largest of three islands] would have to wait about a million years." And even a two-thousand-pounder wouldn't destroy the habitat, though it would make what he called a "hole" and cause local damage. A one-ton meteoroid would average a little less than half a yard across and make a hole twenty-one feet across, O'Neill calculated. He figured that smaller ones, weighing about what a tennis ball weighs, would strike on an average of once every three years and do no serious damage.

O'Neill also made the point that Earth's gravity attracts meteoroids, whereas the islands would have no Earth-like gravity and would therefore not pull in stray cometary fragments and rocks. But he failed to mention that Earth's atmosphere screens out tennis balls, basketballs, beach balls, and their big brothers, including iron asteroids that would go through a space habitat's skin like a bullet through butter. O'Neill also noted that there are three kinds of radiation on Earth, coming from soil, bricks, rocks, and other solids; from inside our own bodies; and from cosmic rays that penetrate the atmosphere. There is half the Earth's radiation in space, he added, but there is a caveat: solar flares. These intense bursts of radiation, which could be harmful to the colonies, are generally considered to be the most dangerous environmental obstacle to living in space for long periods. The solution, O'Neill believed, was to provide the islands with passive shielding—lunar surface material from mining operations, for example—that is perhaps twenty inches thick. He also predicted that later islands—really large ones—would have their own protective atmospheres.

It is the protective atmosphere of the colony called Earth that keeps its dwellers safe from the deadly bursts of X-rays and gamma rays that travel almost at the speed of light. Yet they can affect life beneath the atmosphere anyway. On August 24, 1998, there was an explosion on the Sun as powerful as 100 million hydrogen bombs. Hundreds of working satellites registered the wave of X-rays that suddenly slammed into them. Minutes later a second wave, this one of speeding solar protons, struck. Earth's magnetic field shuddered from the barrage. Ham radio operators had a serious shortwave blackout. And the Sun is by no means the only star whose explosions reach Earth. A few days after the August 24 flare, another barrage of X-rays and gamma rays arrived. They were from a star named SGR 1900+14 that is forty-five thousand light-years away. A nurse in Seattle who was driving home from work at 2 A.M. while listening to a local radio station suddenly heard one in Omaha, and hams chatting on the East Coast picked up voice transmissions from remote places in Canada.

But flare activity is far more serious in orbit and beyond. It takes two words—*solar flare*—to send space-walking astronauts racing back to the safety of their spacecraft. Construction of the *International Space Station* and its successors, repairing and refurbishing orbiting spacecraft, and returning to the Moon become frighteningly dangerous when there are flares. The space agency has therefore been studying them and the other sources of space radiation for years. It operates a Space Radiation Laboratory at the Brookhaven National Laboratory on Long Island in New York and holds a Space Radiation Health Investigators' Workshop every year. And the Universities Space Research Association, which is headquartered in Houston and has nearly a hundred universities as members, is also actively studying the problem. The *Ulysses* spacecraft, whose mission is primarily to orbit the Sun to collect many kinds of information, has sent home useful data on its flares. So has *2001 Mars Odyssey*, which has used its own X-ray and gamma-ray detectors to register hits. Other spacecraft enlisted in solar flares studies have included the *Compton Gamma Ray Observatory*, *Mars Observer*, the *Pioneer Venus Observer*, and *NEAR*. All of them work in a program called the Third Interplanetary Network, which began in 1990. The machines have been, and continue to be, used in combination to calculate exactly where the deadly bursts are located by calibrating the precise time they hit individual spacecraft. And on February 5, 2002, a dedicated satellite called the High Energy Solar Spectroscope Imager, or HESSI, was orbited specifically to take unprecedentedly clear pictures of solar flares' X-ray and gamma-ray bursts. The data will be used to protect space travelers long before O'Neill's huge colonies are built, if they ever are.

O'Neill envisioned the three islands as progressively larger and numbered consecutively. Island One was to be basically a large sphere, topped by huge doughnut-shaped tori in which food would be grown. He made the point that a minimum of a few thousand people would have to occupy the structure if it was to accomplish its two missions: being a self-supporting colony and helping Earth economically. Like the two other islands, the first would be parked at a libration point, so it would be stable for as long as Earth and the Moon stay intact. O'Neill wrote that L5 has come to stand for any stable orbit around Earth that is far enough from both Earth and the Moon so neither of them blocks the sunlight necessary to provide solar power. He inspired the creation of the L5 Society, a group that took space colonization seriously, and which eventually merged with the National Space Institute to form the current National Space Society.

Island Two was conceived as being built with materials mined by a small colony of laborers on the Moon and flung in the island's direction by what O'Neill called a mass driver. This machine, which was conceived by Arthur Clarke and described in his article in the *Journal of the British Interplanetary Society* in 1950, was to be a kind of electromagnetic recirculating conveyor belt that would fling an endless stream of buckets filled with lunar material to the place in L5 where it was needed for construction. That second island would be roughly six thousand feet long, have an equatorial circumference of four miles, and contain 140,000 people, possibly in small villages separated by parks, according to O'Neill, who had fond memories of having once lived in a small Italian hill town.

The largest of the colonies, Island Three, was to be roughly four miles wide, twenty miles long, and have a total land area of five hundred square miles. ("The numbers will seem staggering," he was quick to note, "but they are backed by calculation.") O'Neill's calculation showed that the largest communities in space that could be built with ordinary materials such as iron, aluminum, and special glass, and have oxygen pressure equivalent to five thousand feet above sea level on the home planet, would be eighty miles long, sixteen miles wide, and, as he put it, have a land area half the size of Switzerland.

At any rate, Island Three, like the others, would have artificial gravity, air, land, water, and natural sunshine in an Earth-like environment. The gravity, as on von Braun's doughnut-shaped station and on Islands One and Two, would come from rotation. In this case, O'Neill proposed using two twenty-mile-long, two-mile-wide cylinders built parallel to each other and with ends that were closed with hemispherical caps. The cylinders would rotate independently, simulating gravity for everything inside them. Each cylinder would in turn be divided into three long "valleys" separated by as many arrays of windows. Each valley would be designed primarily as a residential area and have landscaped grounds and parks just like on the home planet. He was well aware that people would insist on a homelike environment in space or they would not voluntarily leave Earth.

Typical of the detail with which he approached the design, O'Neill conceived of mirrors set above the windows so that the valleys would not only be bathed in natural sunshine, but the Sun would appear to be motionless even though the cylinders were rotating. A bowl-shaped mirror at the end of each cylinder would collect sunlight for energy to run the island's (he carefully called it the "community's") power plant twenty-four hours a day. Crops would be grown either in tori or in small cylinders outside the big ones. Crop cylinders, he explained, would accomplish what has never been possible on Earth: "independent control of the best climates for living, for agriculture, and for industry all within a few miles of each other." There would of course be no hurricanes, frosts, floods, or snowstorms. Winter would never come. O'Neill estimated that Island Three could easily support 10 million people, as well as farm animals.

O'Neill wanted people living in space to eat pretty much as they would back home. He learned that cattle are not good at converting plant food to protein-rich meat, but chickens, ducks, and turkeys are. And so, to a lesser extent, are pigs. "With a varied diet including all the corn, cereals, breads, and pastries that many of us enjoy, and with plenty of poultry and pork, the space colonists will have good reason to follow our Pilgrim ancestors, and celebrate Thanksgiving with a feast of turkey, and Christmas with a savory ham," he predicted. "There will be no need for anyone to think in terms of pressed soybean cakes of fish-meal unless they happen to like such things." O'Neill intended earthlings in space to do pretty much everything their counterparts on the home planet do. That, after all, was the point.

Gerard O'Neill was enigmatic. Some believe his minutely detailed blueprint for the colonization of space is deceptive, unworkable, and even shameful. Roger D. Launius and Howard E. McCurdy are among them. Launius was NASA's prolific chief historian for more than ten years before becoming the chairman of the Space History Division at the National Air and Space Museum in Washington. McCurdy is a professor of public affairs at the American University in Washington and a respected space historian. He and Launius are the authors of *Imagining Space: Achievements, Predictions, Possibilities, 1950–2050*, an imaginative and well-documented tribute to the entire space enterprise. In it, they expressed contempt for O'Neill's grandiose "cosmic arks" while showing that the subject is emotionally charged.

"Such fantastic schemes contribute little to the solutions required to sustain the planet. They constitute a form of denial in that they direct public attention to solutions that are technically infeasible. Humans will learn to live on the harsher portions of their own planet—such as polar regions or under the seas—before they move in large numbers to space colonies or Mars. . . . Humans," Launius and McCurdy continued, "had best learn to take care of their own planet before abandoning their home for starry evacuation schemes." Nor were they alone. Others in the space community have simply dismissed O'Neill as being outdated and irrelevant.

But O'Neill is venerated at the Space Studies Institute in Princeton, which he founded in 1977, the year after *The High Frontier* was published. Freeman Dyson wrote the introduction to the third edition of *The High Frontier* while he was president of the Space Studies Institute. In it, he not only praised O'Neill as a visionary, but expressed his own contempt for the agency that is the guardian of the space program. Dyson opened his introduction by stating, matter-of-factly, that O'Neill's vision of humans in space has not been realized, and while he did not set down a definite timetable, he had died believing that at least Island One would be in place by 2010. That, Dyson wrote, will not happen. Instead, he added, there is the *International Space Station*, which in no conceivable way fulfills O'Neill's dream: it will not be commercially profitable; it will not produce anything that justifies its cost; and it will not be spacious or comfortable. Not only is the *ISS* not a step forward, Dyson wrote, it is a big step backward: a setback that will take decades to overcome. "It's no wonder that the general public, seeing the obvious irrelevance of the *International Space Station* to human needs, concludes that O'Neill's dream is equally irrelevant." It is therefore up to those who believe in the dream, Dyson continued, to explain what went wrong.

Freeman Dyson concluded, "Believers should be warned that the infrastructure described by O'Neill in 1976, with launches provided by the Shuttle and other Shuttle-related vehicles, no longer makes sense. Believers must accept the fact that O'Neill's reliance on NASA to bring his dream to reality was a grand illusion. Believers must resign themselves to a long delay, probably as long as fifty years, before the dream will become practical." Ray Erikson, Andrew Turner, and others would counterargue that, whatever its limitations, the *International Space Station* is in orbit and functioning; it is not a pie-in-the-sky concept. They no doubt would also argue that the *ISS* did not do in O'Neill's first island. What did it in was its being a concept too far ahead of its time when there are more pressing priorities.

But Dyson's frustration is understandable, the more so because when he wrote the introduction to *The High Frontier* at the turn of the millennium, he knew about threats to Earth that its author could not have known. One had to do with asteroids and other Near-Earth Objects. As mentioned, O'Neill considered the danger posed by cometary fragments and other objects flying around in space, but only as it related to his islands. He made no mention of the danger NEOs pose to Earth. Failure to do so was not a blunder. He had no way of grasping that particular threat when he wrote his book because the Alvarezes and their two colleagues didn't publish their theory about the Cretaceous-Tertiary extinction event, based on the research in the Yucatán, until 1980. When O'Neill thought about large asteroids, in fact, it was in the context—still a popular and plausible one—of mining them. He even devoted a chapter to homesteading huge rocks the way prospectors scratched for minerals and gems in the Old West. Certainly there was little, if any, thought of Earth being clobbered by one of the big boulders.

Nor had O'Neill any way of anticipating the proliferation of weapons of mass destruction and the means of sending them to their targets. He obviously was aware of the weapons' existence, living as he did in the throes of the Cold War. But he was so focused on the space habitats that even nuclear war was mentioned only in the context of its occurring between opponents in space, noting as he did that people are destined to take their evil sides with them wherever they go, including to Earth orbit and beyond. Certainly the use of weapons of mass destruction by terrorists was not considered because they were tightly controlled by the powerful nations that owned them. Terrorism in O'Neill's time consisted of an occasional airliner hijacking or the kidnapping and murder of a political opponent, not the possibility that megaterrorists would work relentlessly to acquire WMD to commit as much mass murder as possible.

The world has indeed become a dangerous garden, as James Woolsey's poisonous snakes have hatched and spread, resources have been grossly depleted, the environment has been attacked, and civilization's precious records and artifacts have been pillaged or destroyed. But the means to not only survive the collective blight, but in many cases reverse it, is at hand. What is necessary, as Freeman Dyson would say, is the will to overcome inertia and do so.

# : 7 :

# A TREASURE CHEST
# ON THE MOON

A refuge—an island—exists as it has for roughly 4.5 billion years. It has been visited six times by eighteen men from Earth, twelve of whom spent almost three hundred hours studying its surface, and who left behind six of their country's flags, six descent modules, three lunar rovers, five Apollo Lunar Surface Experiment Packages, other science experiments and hardware (including ultra-low-tech hammers), and a photograph of one of their families. Getting them and all that equipment there, and returning the men safely to Earth, required unprecedented scrutiny of the place. And it continues.

The general public had no passion for sending people to the Moon before Aldrin and Armstrong landed on it. As Bruce Murray, the infinitely wise and knowledgeable former Caltech planetary scientist and director of its Jet Propulsion Laboratory, has noted, the Moon has always been a metaphor for the unattainable: "You might as well ask for the Moon. . . . He expects the Moon." And as would have been expected of ordinary people with needs and aspirations closer to home, as mentioned, polls at the time and afterward showed that large numbers of Americans did not favor the voyages or were indifferent to them, and most did not want to pay for them. Interest began to fall off rapidly after the first lunar astronauts came home. But four tribes in the space community remained fixated on sending more people there for a permanent occupation. There was a tiny but vocal minority in the military, chiefly the air force, who wanted to put weapons there. (They were easily overruled by generals who wanted more useful weapons, such as planes and missiles, and who knew the cost of weaponizing the Moon would have been out of the question.) There were scientists who wanted to study the Moon for clues about the creation of Earth and the solar system, and astronomers who knew its far side is an unexcelled place to put an observatory. Entrepreneurs with a commercial interest thought, and continue to think, that mining the lunar surface to supplement Earth's dwindling resources would be both profitable and good for the planet. And there were, and continue to be, the unabashed space enthusiasts—the true believers who joined organizations like the National Space Society, the Planetary Society, and the Mars Society—who were and remain convinced it is humankind's manifest destiny to explore and populate regions beyond Earth and that a lunar colony is the logical place to start the great migration.

As early as 1956, with a number of articles on the subject having already been published by the *Journal of the British Interplanetary Society*, American plans for a manned lunar base were being hatched on both coasts. That September, two Rand Corporation researchers, R. D. Holbrook and H. A. Lang, produced a sketchy "Technical Program Planning Document, Lunar Base Study," which they shared with John E. Arnold, a professor of mechanical engineering at MIT. Arnold taught a course called "Creative Engineering," and as O'Neill did at Princeton with the orbiting-colony concept, he put a hundred freshmen and more than sixty seniors and graduate students on the Moonbase design problem. The most plausible of the senior and grad-student ideas were distilled in five reports that were in turn published by Rand, which contracted with the air force.

With the Cold War well under way, and the Communists having trumped the West by starting the space age in October 1957, a succession of lunar base studies were produced. They came from Rand, the Glenn L. Martin Company, Lockheed Missiles and Space Company, and elsewhere and were read with great interest by a handful of army and air force officers. A few in both services not only considered the possibility of building bases on the Moon, but were haunted by the thought that the secretive and ever-devious Soviets could get there first. That notion had already occurred to Robert A. Heinlein, a prolific and influential writer whose *Stranger in a Strange Land*, *The Moon Is a Harsh Mistress*, and other novels are science fiction classics. He described the possibility of a Communist missile base on the Moon in *Destination Moon*, which was made into the 1951 film of the same name. In it, a retired general named Thayer who is trying to get a privately financed mission to the Moon under way in the face of government indifference warns the executives he is courting, "We're not the only ones who are planning to go there. The race is on and we'd better win it because there is absolutely no way to stop an attack from outer space. The first country that can use the Moon for the launching of missiles will control Earth. That, gentlemen, is the most important military fact of this century." It was a play on the old "high ground" military strategy; in this case, the ultimate high ground. But it was nonsense. It would have taken the missiles at least three days to reach the United States, as opposed to half an hour from Siberia over the north pole, and they would have been spotted by radar and telescopes on launch. The Kremlin and much of the rest of the Soviet Union would have been incinerated long before the missiles reached their targets.

As usual, Clarke was ahead on that one, too. The great visionary concluded six years before *Sputnik* that a military base on the Moon would be counter-productive. "The problem of supply—often difficult enough in *terrestrial* military affairs!—would be so enormous as to cancel any strategic advantages the Moon might have," he wrote in the extraordinarily insightful *The Exploration of Space*, which also appeared in 1951. "If one wants to send an atomic bomb from A to B, both on the Earth's surface, then taking it to the Moon first would be an extremely inefficient procedure. Moreover, a lunar-launched missile could be detected a good deal more easily than one aimed from the other side of the Earth."

The army nevertheless developed a comprehensive lunar-base study called Project Horizon in 1959. It followed by a year an address made at the Aero Club of Washington by an air force brigadier general named Homer A. Boushey, who attained instant notoriety by publicly suggesting that a manned military base on the Moon would provide unparalleled views of Earth and a superlative missile base. The general's remarks were picked up by *U.S. News & World Report*, which ran them under the headline "Who Controls the Moon Controls the Earth." Boushey was ridiculed by Lee A. DuBridge, the president of Caltech, and other scientists. And although he was formally defended by the air force's deputy chief of staff for matériel, many senior air force officers winced at the Flash Gordon image Boushey's plan brought to mind. But Boushey did not speak alone. In April 1960, two years after his remarks at the Aero Club, the air force's Directorate of Space Planning and Analysis came up with a secret plan for a Military Lunar Base Program that called for an underground, self-supporting base to be in operation by June 1969 that would contain an "earth bombardment system" whose missiles would have been able to strike targets on Earth with an accuracy of between two and five miles.

Clarke had already deplored the rocket's use as a weapon that could cause tremendous devastation from wherever it was launched. "It is one of the tragic ironies of our age that the rocket, which could have been the symbol of humanity's aspirations for the stars, has become one of the weapons threatening our civilisation," he wrote in *The Exploration of Space*. He expressed profound sadness because most rocket research at that time was for fighting wars and noted that the technology for civilian and military rockets was the same.

Scientists have had a far greater interest in the Moon than have the airmen, though there have been the usual squabbles over whether people are necessary to do the work, or whether it can be left to cheaper machines. The modern push to send scientists to the Moon began in 1984, partly because President Reagan approved space station *Freedom*, which was taken to be a valuable staging point for a return to the Moon. Furthermore, the Soviet Union's emphasis at the time on long-duration flights on its own stations convinced some members of the space community that the Russians were practicing for a spectacular feat, perhaps starting a lunar base, or even using the Moon to strike out for—what else?—the Red Planet.

Whatever the case, three hundred scientists, engineers, managers, and an occasional Apollo astronaut met at the National Academy of Sciences in Washington in November 1984—fifteen years after Armstrong and Aldrin went to the Moon—to discuss and promote a lunar base for scientific research. The event was sponsored by NASA and was the result of two years of agitation by a pair of committed scientists at the Johnson Space Center, Wendell M. Mendell and Michael Duke, who had earlier won a small grant to conduct lunar base workshops. Their report formed the basis for the meeting in Washington, where it was estimated that the base would cost as much as $90 billion and take as long as twenty-five years to complete, according to then NASA administrator James M. Beggs. One of the scientists who came to the meeting, a geologist from the University of New Mexico named G. Jeffrey Taylor, excitedly announced that lunar scientists in Hawaii had just concluded that the Moon was formed when a planetesimal the size of Mars slammed into Earth more than 4 billion years ago, not long after Earth came into being. The explosion sent billions of tons of debris into orbit that eventually congealed into the Moon, Taylor theorized. This has come to be accepted as the giant-impact, or big-splat, hypothesis. There are no more planetesimals roaming around the solar system, but that horrendous collision serves as a reminder that the universe is a violent place.

As usual, the stumbling blocks for returning to the Moon were funding and factors of politics. James A. Van Allen, who had discovered the doughnut-shaped radiation belts that bear his name with a radiation sensor on *Explorer 1* twenty-six years earlier, was an implacable opponent of manned spaceflight because he believed it severely crippled scientific research, which is the only real justification for a space program. "I would be delighted to have a station on the Moon," he said. It would "obviously fulfill a long-felt desire of certain regions of humanity." But talking about it when many areas of science were lacking adequate financial support, Van Allen added in an apparent reference to the old wives' tale, "is cheese in the sky."

But the astronomers, who had long dreamed of getting their telescopes on the Moon, did not agree. They argued that observations and experiments could be done on the lunar surface that are impossible on Earth and more difficult from orbiting stations. (The Great Observatories, one of which was to be Hubble, came later.) Bernard F. Burke, a leading astronomer, proposed arrays of optical telescopes spaced many miles apart on the lunar surface that would produce far more detailed images than are possible from Earth, giving "an absolutely revolutionary new view of objects in the universe."

"Most of the arguments presented at the Academy—that we should go back to the Moon to study the origin of the solar system, to build giant observatories on the far side, to exploit lunar resources, or mount further expeditions to Mars and the asteroids—have been made many times before," the journal *Science* reported. Yet the participants "formed a solid and responsible cross section of the space community, and they were convinced that the time had come to take the idea seriously. 'We see the Moon as a logical, evolutionary step,' Mendell said." Yet with nothing more pressing than science to drive it, the prospect of returning to the Moon—even with the tenuous blessings of the Reagan White House—ultimately came to nothing.

Neither did the prospect of turning the lunar landscape into a commercial enterprise, which has run through the literature, technical and popular, from the beginning. The authors of the *Collier's* series in the early 1950s took on the question of the ownership of the Moon and the planets, partly in the context of mining and other resource exploitation, and concluded that an analogy could be made with Earth's oceans.

That is, the Moon itself would be the common property of all humanity, but resources taken from it, like pearls, oysters, and sponge beds in "sedentary fisheries" on Earth, would belong to the nation that takes them. The notion of "living off the land" on the Moon was mentioned in *The Exploration of Space* and was the subject of a Lunar and Planetary Science Conference in Houston in April 1983. Liquid oxygen manufactured from lunar soil on the Moon itself could be used as rocket propellant for space operations, it was concluded, possibly leading to the creation of Moon base in the late 1990s.

And the commercial rationale for returning to the Moon continued. A series of respected reports prepared under the auspices of the federal government between 1986 and 1991 declared that resource exploitation was integral to the establishment of a lunar base. "Pioneering the Space Frontier," a 1986 study in which the National Commission on Space tried to set a civilian space agenda for the twenty-first century, explained that prospectors would comb the lunar surface, "as is customary in resource exploration." "Leadership and America's Future in Space," the so-called Ride Report, after Sally K. Ride, its author, was a NASA-sponsored document that described an outpost on the Moon in altruistic terms alluding to the manifest-destiny philosophy. So did the National Commission on Space report. Both came out right after the loss of *Challenger*, when a traumatized nation needed an uplifting goal in space. Ride, the first American woman in space, explicitly called for the creation of "lunar mining and processing technologies to enable the prospecting for lunar resources." Four years later, the Synthesis Group, a number of experts headed by General Thomas P. Stafford, another former astronaut, produced a report called "America at the Threshold," which similarly tried to chart the nation's course in space. It called the Moon "a source of both materials and energy for an emerging space-based economy in the 21$^{st}$ century."

It takes nothing away from the Synthesis Group's report to note that it was formally submitted to Vice President J. Danforth Quayle, who was also the chairman of the National Space Council. In that capacity, he electrified the science community on August 18, 1989, when he said, straight-faced, in an interview on the Cable News Network, that humans could thrive on Mars because "we have seen pictures where there are canals, we believe, and water. If there is water, that means there is oxygen. If oxygen, that means we can breathe." And where earthlings can breathe, the vice president assured the interviewer, they can function quite nicely. It was one of Quayle's more memorable gaffes.

Prestigious groups such as the Space Frontier Foundation and the Space Studies Institute held and continue to hold regular meetings that address the colonization and development of the Moon, as well as broader issues. The foundation's goal is to transform space from a "government-owned bureaucratic program into a dynamic and inclusive frontier open to people" and to start settlements in space as soon as possible. With the Moon as a priority candidate for one or more settlements, the foundation regularly holds a Return to the Moon conference at which government, academic, and industry representatives who specialize in space-related activities make presentations.

The Space Studies Institute has been holding biannual conferences on space manufacturing since Gerard O'Neill was active there. The three-day meeting in July 2001, the thirteenth such conference, had sessions on space power and energy; robotics; living in space; transportation; asteroids and manufacturing; and international, legal, and economic considerations of settling the Moon; plus an evening roundtable discussion. Papers presented in the living-in-space session addressed topics such as "Lunar Base Development: Design Aspects, Technology Requirements and Research Needs"; "Sustainable Water in Closed Biosystems: Preliminary Design Considerations"; "Electromagnetic Construction of a 1km-Radius Radiation Shield"; and "A Rationale for a Human Lunar Pathfinder Mission." One presentation concerned "Doing Business in Space: This Isn't Your Father's (or Mother's) Space Program Anymore." As at the other meetings, all the presentations, in this case thirty-two, were bound in a hardcover book, replete with equations and complicated diagrams, this one titled *Space Manufacturing 13: Settling Circumsolar Space*. It did not make for light reading, but it was filled with the kind of highly specialized, functional material that by then existed in many areas relating to space, and certainly to the Moon.

In 1987, the year Sally Ride's report appeared, John S. Lewis and Ruth A. Lewis published *Space Resources: Breaking the Bonds of Earth*, which took a hard but optimistic approach to gathering minerals and other resources on the Moon, Mars, asteroids, and elsewhere. It also addressed manufacturing commodities on the planets (including the Moon) and in low Earth orbit on the then planned space station. The lead author was a professor of planetary science at the University of Arizona who was widely respected in the space community. He certainly understood the incredible difficulty of just returning to the Moon, let alone mining it for resources and doing manufacturing, then reaching much farther out to Mars, and mining it and asteroids as well. What he and other responsible scientists wanted to do was explain how those things could be done, that they were technically feasible and would someday be fundamentally important to civilization, much as O'Neill had.

Nor did *Space Resources* describe the only clearheaded, richly detailed plan to exploit space. The *Journal of the British Interplanetary Society*, which is not known to be recklessly altruistic, devoted its entire February 1997 issue to the "Economics of Space Commercialization." There was an article by John Lewis on mining asteroids, and others on space tourism, space hotels, and creating an infrastructure for commercial space. The last contained a section called "Access to Space," which was emblematic of what finally went wrong with the plan to return to the Moon:

"Programs such as the United States' National Aeronautics and Space Administration's (NASA) X-33 and X-34 programs—which aim to produce reusable spacecraft for large and small payloads respectively—have begun to pursue the concept of Single Stage to Orbit (SSTO) spacecraft as a means of reducing the cost of transporting mass to orbit. Although managed by NASA, these programs are being developed in an industry partnership with the intention of producing and operating a spacecraft financed by the private sector." But the private sector could not sustain either project because of a serious lack of financial support. That, in turn, was because there was nothing about the idea that was sufficiently compelling where Congress, its constituents, and the White House were concerned. Both spacecraft, as noted, died on the drawing board. The proverbial bottom line was that the whole Single Stage to Orbit and Two Stage to Orbit ideas carried more risks than benefits, given their missions, which were poorly defined, badly articulated, and less than compelling relative to the cost of implementing them.

Indeed, talk has been a good deal cheaper than action. The space crowd became elated on July 20, 1989, for example, when President George Bush took the occasion of the twentieth anniversary of the Apollo 11 landing to proclaim on the steps of the National Air and Space Museum the birth of a Space Exploration Initiative. The SEI, he explained, was to be an audacious long-term commitment to send Americans back to the Moon and on to Mars for a permanent presence in space. "What was once improbable," Bush announced, "is now inevitable." It was sheer political theater, an expedient, vacant gesture. The exploration initiative had no serious plan, no timetable, or anything else to suggest the White House was taking it seriously. But the White House, probably trying to echo and even trump JFK with a glamorous, ennobling long-term program, had asked the space agency for a goal. And it got one.

NASA then scrambled to come up with a plan, something concrete, that would make the SEI substantive. Three months later, it produced a report that contained a rough, thirty-year timetable for establishing the base on the Moon and reaching Mars. The document leaned heavily on the Ride report, which had come out two years earlier, and which had specifically referred to a Solar System Exploration Initiative encompassing both manned and robotic missions. The rumored cost, which was never officially estimated, was $400 billion. That made even the most dedicated space enthusiasts in Washington blanch. Worse, the document was loaded with mission descriptions and hardware requirements, but contained no compelling reason with which the initiative could be sold to the public and Congress. It was the equivalent of Sir Edmund Hillary's famous justification for climbing to the top of Mt. Everest: "Because it is there." But it didn't cost $400 billion to reach the top of the mountain. Bush later told Sean O'Keefe, who became NASA's tenth administrator in 2001, that he had been "set up" by the space agency and its amorphous wish list.

"In hindsight," Mendell said, "NASA leadership should have asked for more specific guidance before preparing their report. They took the president at his word. I talked to people very high in NASA who really believed that if the president wanted to go to the Moon and go to Mars, that he could get the money. I just don't think it works that way."

Then there was the looming threat from Asia, at least as it was perceived by a stalled U.S. manned lunar program. Starting in 1996, plans were made known that Japan—the "Land of the Rising Moon," as *Space News* put it—intended to start a lunar program. The idea, as announced by Japan's Lunar and Planetary Society at a workshop in Kyoto that October, was to send three Japanese to a Japanese lunar base every six months starting in 2024. Each crew would stay for a year. Hidehiko Mori, an engineering adviser with the National Space Development Agency of Japan, told delegates to the workshop that the development of less expensive large launch vehicles was the key to Moon-base construction, and that the development of the Single Stage to Orbit lifter, which he claimed was ten times cheaper to launch than the shuttle, was vital to the program. The Lunar and Planetary Society estimated it would cost $26.1 billion to build a Moon base within thirty years with international cooperation. It is often forgotten that Japan used its huge H-2 heavy-lifter to get two small orbiters to the Moon in 1990, making it only the third nation to reach it.

India and China followed suit. Both have also announced plans to reach the Moon. Indian scientists announced in the spring of 2003 that they supported their nation's first venture beyond Earth orbit: a proposed $100 million effort to place a five-hundred-pound spacecraft in a sixty-mile-high orbit around the Moon. And the chairman of the Indian Space Research Organization provided a suitable sound bite: "It is not a question of whether we can afford [to go to the Moon]. It is whether we can afford to ignore it." Not to be outdone, the People's Republic of China weighed in with plans to send its own spacecraft to the Moon to conduct still more scientific experiments. And as the director of the country's National Space Administration blithely added, another goal is to exploit resources such as helium-3, a reasonably safe nuclear fuel that has a relatively low neutron flux that could be used in fusion reactors on Earth (if any ever operate). China, however, took what is being called the new space race a great leap forward by orbiting Yang Liwei. And there is much more to come. Not continuing the manned space program would make as much sense as calling off the Soviet and American programs after Gagarin's and Glenn's flights. China has made it clear that the route the two Cold War superpowers took to space is precisely the one it will take. That means missions with two or more men or women, construction of a station, and sending a robot lander to the Moon by 2010. The lunar lander would be a precursor to sending humans.

There are two fundamental reasons the People's Republic of China, as well as Japan, India, Brazil, and other nations, wants an ambitious presence in space: political prestige and military capability. A large space program, and certainly one that is able to send robot explorers to the Moon and people and stations to orbit, requires a huge, advanced scientific and technological base that suggests a stable and powerful political system; what used to be called national "might." That, in turn, fosters international power. Being a space power, particularly by being able to send citizens there, has become as important for nations with regional or global ambitions as developing nuclear weapons. Military applications, in fact, figure decisively in space. When Yang Liwei went into orbit, the *People's Liberation Army Daily* ran a story predicting that space would become a "sphere of warfare" because space-based satellite technologies are critical for a modern military. The use of satellite technology, certainly including reconnaissance satellites and satellite-guided smart bombs, to pulverize the Taliban in Afghanistan was not wasted on the Chinese military. Nor was the political value of heading for the Moon. Beijing has been explicit about sending two astronauts to space on five- to seven-day missions and eventually going on to the Moon. Seven months before that first manned flight, Luan Enjie, the director of the China National Aerospace Administration, was blunt in saying that the Moon is "the focal point wherein future aerospace powers contend for strategic resources." His and other officials' calls for mining the Moon for helium-3 and other resources caused some American experts to warn darkly of a "Red Moon" with a manned Communist base. It was General Thayer redux.

With that and other possibilities in mind, and talk of a new space race in the air, the National Security Council drafted an abstract plan that called for returning Americans to the Moon by 2020 and then going on to Mars at an unspecified time. That was the genesis of the "space initiative" announced by President Bush on January 14, 2004.

Anticipating that the rest of the world would interpret the American plan as a continuation of the White House's unilateral foreign policy, this time to occupy the Moon, those who wrote Bush's speech had him say, "The vision I outline today is a journey, not a race, and I call on other nations to join us on this journey in a spirit of cooperation and friendship." Yet, invoking the spirit of Lewis and Clark, he referred to the nation's pathbreaking history in space and said it was "time for America to take the next step." The only justification for returning to the Moon Bush mentioned in the vague plan was turning its soil into rocket fuel and using its relatively low gravity to facilitate going to Mars and elsewhere. "We do not know where this journey will end," he said solemnly, "yet we know this: human beings are headed into the cosmos." Precise reasons for sending people back to the Moon and beyond, technological requirements, and other details were left to be worked out by a specially created presidential commission.

The major sticking point was immediately raised: money. As noted, Bush called for completing the *International Space Station* with the remaining shuttles by 2010 and then retiring them. He also said he would direct the space agency to divert $11 billion from existing programs, especially from the shuttle's five-year budget, and ask for an additional $1 billion over those five years. That was as specific as he got about financing the return to the Moon. Two weeks later, the Congressional Budget Office predicted that the federal budget deficit would hit a record $477 billion in 2004 and accumulated deficits during the following decade would set another record: $1.9 trillion. By coincidence, 2014, which would mark the end of that decade, was the year mentioned by Bush when the otherwise undefined Crew Exploration Vehicle was to have been launched.

Some thought the initiative was a bold and imaginary way of invigorating a listless space program and rekindling the nation's sense of adventure. Others thought it was merely an election-year feel-good political stunt by an administration stalked by an imbroglio in Iraq, a weak economy, and other problems. And still others saw merit in "the vision thing," but were dubious about its being achieved. They all squared off within hours of the announcement. The American Institute of Aeronautics and Astronautics—the same professional group that had alerted Congress to the asteroid threat a decade earlier—issued a news release strongly endorsing Bush's call for an "exciting and challenging space mission": "President Bush celebrated the end of America's First Century in Flight by proclaiming a bold mission that marks humankind's commitment to slip the surly bonds of earth and to develop a suite of technologies that will permit us to inhabit another planetary body—first the Moon—and eventually Mars." John Bahcall, a physicist at the Institute for Advanced Study in Princeton and a longtime NASA adviser, called the space initiative inspirational. "The whole agency is challenged to refocus its efforts. Everything will be turned upside down. It will be an exciting place to be." But two days later, it was space science that took the first hit after the president's announcement. Astronomers and managers at the Space Telescope Science Institute in Maryland were given some devastating news by Sean O'Keefe, who told them there would be no more servicing missions to the Hubble Space Telescope by shuttles. The official reason was that shuttle servicing missions were excessively dangerous because the orbiters didn't carry enough fuel to make it to the space station in case of trouble. O'Keefe reemphasized that point in Senate testimony and in a letter to Senator Barbara A. Mikulski of Maryland. He categorically denied any motive other than safety, saying the total risk "is too high." But others suspected the real reason for canceling the flight was to cut costs to help pay for the new initiative. Each service mission cost $500 million. Whatever the reason, it was a death sentence for the crown jewel of astronomy, a truly wondrous machine that could see back toward the beginning of time. Dr. Steven Beckwith, the director of the Space Telescope Institute, said Hubble could "die tomorrow" or last until 2007. "We at the institute are devastated by the potential loss of Hubble. But we will do our absolute best to make the final years of its life the most glorious science you've ever seen." Dr. Tod Lauer, an astronomer in Tucson, was openly bitter: "This is a pretty nasty turn of events, coming immediately on the heels of W's endorsement of space exploration," he said.

Howard McCurdy, the historian, told *The New York Times* that the space initiative was feasible, but that the odds did not favor it. "Doing it successfully would mean a total transformation of the space program, as radical as happened from 1961 to 1966. For thirty years NASA hasn't had that discipline." John E. Pike, the director of a Washington research group on military and space topics and a longtime observer of the military and civilian space programs, said the new initiative was unlikely to go anywhere. Unlike Kennedy's call for excellence during the Cold War, Pike said, Bush's proposal addressed no underlying national political issue. He called it election-year grandstanding. "The trivial budget increases they're proposing are only going to produce artwork," Pike continued. With shuttles ordered grounded by 2010, and the

Crew Exploration Vehicle only a theory, Pike was deeply pessimistic. "Basically, they looked at piloted space and said, 'Let's shut it down and let's have a hedge against the possibility that the Chinese will go to the Moon.' That's it. There's nothing to replace shuttle and station except artwork."

The most persistent rationale for living on the Moon has been—and remains, for lack of a better word—spiritual. That is not to say religious. Rather, it defines humans as an endlessly inquisitive species whose curiosity has driven them to explore this world and must now send them to others, the first being ones they create themselves—stations in space—and then the Moon and Mars. Sally Ride, who teaches physics at the University of California at San Diego, mixed practicality with the ethereal. "Beginning with robotic exploration in the 1990s," she said in her report, "this [Solar System Exploration Initiative] would land astronauts on the lunar surface in the year 2000, to construct an outpost that would evolve in size and capability and would be a vital, visible extension of our capabilities and our vision. . . . This initiative represents a sustained commitment to learn to live and work in space. As our experience and capabilities on the lunar surface grow, this extraterrestrial outpost will gradually become less and less dependent on the supply line to Earth." She also made the point, now haunting the space program, that a "heavy launch vehicle and a healthy Space Shuttle fleet" are required. "The transfer of both cargo and crew from the Space Station to lunar orbit requires the development of a reusable space transfer vehicle. This and the heavy-lift launch vehicle will be the workhorses of the Lunar initiative."

Planetary defense, including the idea of using the Moon to protect Earth and provide a lifeboat for the people on it, was not on most people's lists of compelling reasons to return. But there were some notable exceptions. The Commission on the Future of the United States Aerospace Industry, a blue-ribbon panel that included Buzz Aldrin and Neil Tyson, issued a final report to the president and Congress in November 2002 that was unique in several respects, not the least of which was because it addressed planetary defense. While it made no mention of going back to the Moon, the thick document did note that the U.S. Air Force was developing a constellation of satellites designed to detect and track other satellites in orbit. The commission suggested that studies for the proposed satellite detection program be broadened to include spotting potentially dangerous asteroids as well (as the Ballistic Missile Early Warning satellites in the Pentagon's Defense Support Program had been doing for years). It went on to note that the U.S. Strategic Command, which oversees the military big picture, was considering establishing a clearinghouse that would collect and analyze data on potential Earth impactors, while space- and ground-based surveillance systems were being integrated. "Given these actions," the commission wisely suggested, "planetary defense should be assigned to DoD [Department of Defense] in cooperation with NASA. The Commission believes that the nation needs a joint civil and military initiative to develop a core space infrastructure that will address emerging national needs for military use and planetary defense."

Recognizing the seriousness of the threat from impactors and invoking planetary defense to protect against them was a breakthrough for that kind of group. But recommending that civilians and the military work together to establish an integrated, joint defense was a true milestone. NASA's Spaceguard, its international counterparts, and the air force's Space Command are complementary entities, one to spot threats and the other to nullify them.

But Andrew Chaikin and Paul D. Lowman Jr. expressed concern about other, more pernicious threats that seem to be unavoidable. They believe that establishing a permanent colony or several of them on the Moon is the only rational course for this race. It would certainly give the Bush space initiative a clearly defined, deeply serious purpose. Chaikin is a talented science writer who is best known for his expertise on the Apollo Program and for a highly regarded book, *A Man on the Moon*, that chronicled it. Lowman is a planetary scientist at the Goddard Space Flight Center who has a long-standing interest in studying and settling the Moon and has written a prodigious number of scholarly papers, using both broad strokes and narrow ones, that explain why it should be done.

Chaikin's reason for colonizing the Moon echoes Gerard O'Neill and the Synthesis Group: human expansion into the solar system is imperative because Earth's resources are finite and its ecosphere is increasingly under siege. He explained in an article in *Air & Space/Smithsonian* that the twenty-seven members of the Synthesis Group were both older veterans of the Apollo era and younger specialists. "Regardless of age, everyone sensed the importance of the task—to find a convincing rationale for the long and expensive effort of human expansion into the solar system. One SG scientist recalls, 'The constant feeling among the members during the deliberations was, "If we lose the chance now, we've lost it for our generation, and we might as well not do any more."' At first, some of the group's senior members resisted the lunar base, fearing it would siphon off resources needed to go to Mars," Chaikin reported. "'Why should we go back?' they said. 'We've already been there.' To which the younger members responded, '*You've* been there. We haven't.'"

Chaikin went on to explain that the Synthesis Group finally endorsed the Moon, not Mars, for the same reason von Braun had four decades earlier: Mars is too far and difficult without going back to the Moon first. Then, fine-tuning the argument, Chaikin invoked the economic and environmental situation in detail: "If the world's population continues to grow at the present rate, the year 2050 will find the planet crowded with ten billion people. They will face a global environmental catastrophe and an energy crisis of staggering proportions if the current dependence on fossil fuel goes unchecked. 'Very few people appreciate what a desperate situation the Earth's in right now,'" Chaikin quoted Maxime Faget as warning. Faget, a legend in his own right, was NASA's chief manned-spacecraft designer and the guiding light behind the Mercury Program and those that followed. He conceptualized Mercury before there was a NASA.

Lowman, who has written copiously about the Moon from the wide perspective of space exploration in general, thinks of the Moon as a fundamentally important stop on the way to Mars, as a valuable resource, and, with Mars, as a lifeboat. He believes the lunar and Mars programs are complementary, not competitive, and have to do with what he calls "dispersal." "Both can be justified in the most fundamental terms as beginning the dispersal of the human species against uncontrollable natural disasters, cometary or asteroidal impacts in particular, to which mankind is vulnerable while confined to a single planet," he told an audience at the Colorado School of Mines in late October 1999. Paul Lowman was sounding the same warning as Captain Osepok in *Encounter with Tiber*: spread out or die.

And Lowman used the long, fruitless search for extraterrestrial intelligence, commonly called SETI, to help make his point. It is possible SETI has turned up nothing since it began in 1960 because the radio astronomy and other ways of trying to tune in to transmissions from distant regions of the cosmos are not being done correctly. It is also possible, Lowman explained in Colorado, that extraterrestrial intelligence is rarer than we think. He was in good company. The authors of a book called *Rare Earth*, which was published in 2000, made the point that there is a great deal more to calculating the chances of complex life being out there than just probability theory. In other words, sheer mathematical probability—if there are a billion stars in this galaxy and intelligent life exists near one of them, then there is probably intelligent life near another star in another galaxy with a billion stars—is simplistic. They made the point that where complex life is concerned, to borrow an old cliché, position is everything. That is, conditions on Earth were just right for the fostering of the life-forms on it; one planet closer to the Sun and it would have been too hot, and one farther out and it would have been too cold. Their theory, which is rational, seems borne out by the fact that more than a hundred planets have now been discovered orbiting suns in this galaxy, but none seem capable of fostering life, at least as it is commonly defined.

Lowman has also warned that intelligent life can extinguish itself. That is, it is possible SETI has turned up nothing because one or more civilizations got too smart for their own good and in effect committed suicide. That possibility has been obvious since Hiroshima, he said, but it has taken an even more serious turn with the discovery in recent years that the human presence on Earth is significantly changing the planet's fundamental environment in ways that are increasingly dangerous. Global warming is an obvious example. So are pollution and natural resource depletion.

"The point of this discussion, in context of a return to the Moon, is that our species is vulnerable to uncontrollable natural hazards as long as we are confined to one planet. Like the cessation of radio signals from the sinking *Titanic*, the absence of ETI [Extra Terrestrial Intelligence] signals may have ominous implications. Dispersal of the species," Lowman warned again, "is becoming realized as absolutely necessary if mankind is to have a long future."

Lowman took the occasion of his talk in Colorado not only to promote lunar science, including astronomy, and the need to disperse, but to challenge the long-cherished idea that the Moon can be mined the way Earth has been. Exploiting the Moon for minerals that would make being there economically feasible has long been a staple of the return-to-the-Moon crowd. But there are no minerals likely to be found on the Moon that could qualify as "ore," in the sense metal could be extracted for profit if it has to be shipped to Earth, he said. That's because the Moon has had a much simpler and shorter geologic evolution than Earth. But there is a single, notable exception: helium-3. If fusion power reactors ever do come online, as many say they must to reduce fossil fuel dependence, the equivalent of one shuttleload of helium-3 is estimated to be able to produce an enormous amount of energy.

As Lowman himself well knew, he followed a long and distinguished line of lunar-colonization advocates, starting with science fiction writers and continuing to other scientists themselves. The British Interplanetary Society, which was founded in 1933, was preoccupied from the beginning with solving the problems of landing on the Moon. The result was the first detailed study of a lunar space vehicle, which appeared in 1939. P. E. Cleator, the society's president, predicted that in spite of the extreme environmental conditions on the Moon "it seems likely that an outpost will be established there." Articles by society members on the intricacies of setting up a base on the Moon exploded in the 1950s. Clarke's *Exploration of Space* appeared in 1951, followed by the *Collier's* series and books that were spun off from it, including *The Conquest of the Moon* by Ley, Whipple, and von Braun in 1953. And there were several others, including the army bureaucrats and engineers who dreamed up Project Horizon.

Clarke returned to the Moon in 1968 in a chapter in *The Promise of Space* that was devoted to a permanent settlement there. In it, he tackled the pesky problem of the resource situation in his usual, delightful way. The most valuable resource of all, he wrote, is water. "It certainly exists on the Moon; the question is where, and in what form." He ruled out water in its liquid form, at least on the surface, because it is too cold. The Apollo landings that began the following year seem to have substantiated Clarke's assumption. But he suggested there could be water ice underground, perhaps in caves where solar heat never penetrates. "Since water is ninety percent oxygen, the two major necessities of life would be provided," he continued with characteristic optimism. "But the hydrogen would be almost equally important, since this is the best of all rocket fuels. Once it could be liquefied and stored, the economics of Earth-Moon space transportation would be revolutionized." Then there was food. "Perhaps by the time (around the turn of the century?) we are planning extensive lunar colonization, the chemists may be able to synthesize any desired food from such basics as lime, phosphates, carbon dioxide, ammonia, water. In fact, this could be done now if expense was no object; it will *have* to be done *economically*, within the next few decades, to feed Earth's exploding population." From there, the unendingly optimistic Clarke launched into a typical rhapsody that foresaw hydroponic farming, habitats ranging from temporary inflatable "igloos" housing a few score scientists to millions of people living in comfortable, totally enclosed cities, and nuclear-powered rockets to connect them to the home planet.

Clarke also came out for reusable launch vehicles, not expendables like the Saturn V, to settle the Moon because, he reasoned, they would be less expensive. "An Atlantic liner that delivered three passengers and sank after its maiden voyage would not be an engineering achievement of which one could be very proud," he has noted wryly. Andrew Turner is certainly in favor of using reusable, or partly reusable, launch vehicles (such as his Aquarius). But he has made the interesting point that a Saturn V could get 250,000 pounds to orbit, or roughly four times as much as a shuttle. In terms of tonnage, two Saturn V flights a year would therefore have carried the equivalent tonnage of eight shuttle flights. "If we had simply built and flown two Saturn Vs a year from, say, 1973 to the present day instead of developing the Space Shuttle, we might have a lunar base and a much larger American Space Station by now." Furthermore, as the experience with shuttles has shown, reusables are only cheaper to use than expendables when they are used a lot. The huge cost of maintaining an Atlantic liner that carried only three people five times a year is a better analogy.

Clarke envisioned vast improvements in launch vehicles to support his lunar colony. In the long term, these may be necessary, but in the short term it is possible to do the early work on the colony using existing technology. As early as 1945, Clarke foresaw today's use of geosynchronous orbits for communication and other purposes with "rocket stations" that would have human crews. But electronics and other technologies advanced faster than he expected. So telecommunication is carried out from the Clarke orbit by robot satellites with lifetimes of fifteen years or more that weigh only a few tons. An early lunar colony could, similarly, be an entirely automated operation that would back up important data on Earth, as will be shown, do meteorological work, and perform more functions done by other "satellites" closer to home. Such a colony would be supported by rovers, which could be put to work constructing facilities and gathering resources in advance of the arrival of a human landing party. The modular equipment could be sent ahead in relatively small batches, then connected by the rovers. In Turner's opinion, such an automated colony could be visited from time to time by astronauts to install or service complex equipment. Eventually, a full-time human presence would become economically justified as the value of the lunar assets increases, just as full-time repair capabilities are justified for facilities on Earth.

There was no colony on the Moon when the millennium changed because the need to start one had been subverted by politicians who (correctly) saw no constituency. But the more farsighted and thoughtful among them could have begun to educate the public about the danger so a constituency could begin to form. That should now be done.

Scientists and engineers in and out of NASA, supported by lawyers, managers, and others who believed spreading out was imperative, had in the meantime taught themselves how to make a colony, and they had done it in painstaking detail. The Moon has been under close investigation by a long succession of American and Russian landers and orbiters that have sent home a vast amount of information on its exact size, composition, gravitational and magnetic fields, and other important characteristics. *Luna 1*, a Russian flyby, started the close-up investigation as early as January 1959 in the initial phase of the race to the Moon. Nine months later, *Luna 3* scored a historic breakthrough when it sent back the first pictures of the Moon's far side. *Luna 10* made the first successful soft landing on the lunar surface in early April 1966. Two months later an American craft, the insect-shaped *Surveyor 1*, helped pave the way for the Apollo astronauts by soft-landing and returning more than ten thousand pictures. It was followed by *Lunar Orbiter 1*, the first U.S. spacecraft to circle the Moon, which also provided a heavy data return. Many sensor-loaded spacecraft followed those early ones, and so, of course, did twenty-one men from Earth.

But sizing up the Moon, both for the sake of pure science and for a base, continued after Apollo and has not stopped. *Clementine*, which was built by the U.S. navy and launched in January 1994, for example, spent two and a half months intensively scrutinizing the Moon. It sent home the first digital maps showing the composition of the lunar surface and stunningly precise topographical maps, in addition to high-resolution thermal imagery. Almost four years later, *Lunar Prospector* sent back a stream of data on the composition of the Moon's surface and the intensity of its gravitational and magnetic fields. It also found evidence that seemed to confirm earlier sightings indicating there was indeed water ice at the poles. And years before either *Clementine* or *Lunar Prospector* got there, their predecessors measured the Moon to the point where there was precise data on its mass, mean radius, surface area, flattening, average density, gravity at the equator, and the escape velocity a rocket would have to develop at the equator to get off the Moon (1.285 miles a second to be precise).

By the end of the millennium, there was a staggering amount of material on the geology of the Moon. There is a library on the rocks and minerals spread over its surface, many of them fragmented chunks of asteroids that impacted and exploded. There is a vast amount of data on the basic composition of the surface elements; soil density, porosity, and thermal and electromagnetic properties; surface temperature and its "lighting environment"; the atmosphere (such as it is); the number of meteoroid impacts (like on Earth, many made by rocks that were too small to be called asteroids or comets); and a great deal more.

The location of a lunar base has also been given a lot of serious thought.

Vladislav V. Shevchenko, a leading physicist and mathematician at Moscow University who has long experience in the Russian space science program, has made the obvious point that the base's location depends on its purpose. However, he adds realistically, "At the first stage of development of a lunar infrastructure, it will be necessary to decide the insoluble task: to combine all various requirements and choose a place for the first lunar outpost." Shevchenko has mentioned, as one example among several, an area in the western part, Oceanus Procellarum, which is northwest of the crater Reiner. It is flat, near unique geological formations, appears to be close to the track made by a recent comet hit, and seems to be in an area of lava streams and a new crater named for another Russian, the enigmatic rocket-engine designer Valentin P. Glushko. Shevchenko has further noted that the eastern part the Reiner crater would be a fine place to put a radio telescope. Robert Goddard, who successfully fired the first liquid-fueled rocket in 1926, favored a crater at or near the poles, also because of the likely availability of water ice.

Shevchenko has warned against carrying some bad habits to the Moon. "Taking into consideration the sad experience of partial destruction by mankind of the Earth's natural environment, it is necessary to take into account conservation aspects in the early planning of Moon exploration. Every location on the lunar surface is a unique witness to the processes of creation and evolution of the solar system and the activity of the Sun on a geological time scale," he has noted. Site selection and building the first base should therefore be planned to avoid destroying the unique places that will be the visitors' new home.

Others, perhaps foremost among them Wendell Mendell, have put a great deal of time into thinking about what the base would need to function; what the engineers call its "elements." First there would have to be a construction shack from which the base would be started. Other elements would include laboratories, a control center for communication and data handling, and habitats, including living quarters, sleeping areas, a place to prepare and eat food, plus laundry, medical, surface-transportation, and recreational facilities. There would also have to be interface modules; air locks; systems to control air temperature, power supply, communication, and life support systems; shielding; and facilities for science, maintenance and repair, storage, and launchings and landings. The base would also need dedicated facilities for mining, which would include places to keep soil movers, drills, and excavation equipment. And there would have to be places for chemical processing, mechanical operations—including furnaces, presses, and machine tools—as well as an electrical shop. A biological production facility that would grow vegetables, meat, and flowers would be on the base, as well as an area for tools and mechanical shops.

Still more thought has gone into detailed specifics on the base's inhabitants. That is, on how to keep them safe, comfortable, and productive. That directly affects the design of their habitats, laboratories, and the air locks connecting them, as well as the thermal control system, power supply and energy storage, communication and navigation systems, shielding requirements to protect them against solar flares and other deadly radiation, and extravehicular activity—in this case, working outside the base's structure, which would require lunar space suits that have also been conceptualized. So have surface vehicles whose ancestors go back to Apollo.

Thinking about the base's life support system, including keeping its occupants healthy, has produced detailed studies on the design of that system, physiological and medical requirements and constraints, physicochemical life support subsystems, and bioregenerative life support subsystems. (The last is engineering speak for producing fresh food.) This would, in turn, involve what the base's planners call the biological life support system. The BLSS would essentially be a self-contained system in which human waste would be recycled and fed to plants, along with the carbon dioxide and water vapor humans exhale. The fruit and vegetables would repay the favor by supplying the people with food and oxygen.

All of this and a great deal more is in *The Lunar Base Handbook*, an excellent 851-page manual on the subject edited by Peter Eckart, a German who has a degree in aerospace engineering from the Munich Technical University and who wrote his doctoral thesis on lunar base design at the Johnson Space Center. Fifty-three international experts, including Arthur Clarke, Buzz Aldrin and other astronauts, and many top scientists and engineers from around the world, contributed to the book, which by its very existence speaks to the imperative of establishing a settlement on the Moon for the safety of humanity.

And civilization is not the only thing to be saved from destruction. There is also its fragile record. The pages of history, or at any rate what remains of them, clearly show that the record of the collective civilization that has inhabited this planet since humans started painting in caves is subject to destruction and loss with no possibility of recovery.

The storied Library of Alexandria comes to mind. The city bearing his name was created on command of Alexander the Great, who encouraged respect for all cultures and the pursuit of knowledge for its own sake, and who wanted a true metropolis that embodied his philosophy. No wonder the library and an associated museum there became the glory of what was then the greatest city on Earth, the first true research institute in the history of the world, as Carl Sagan put it. "Here was a community of scholars, exploring physics, literature, medicine, astronomy, geography, philosophy, mathematics, biology, and engineering," Sagan wrote. "Science and scholarship had come of age. Genius flourished there. The Alexandrian Library is where we humans first collected, seriously and systematically, the knowledge of the world." While working in the library, Herophilus identified the brain rather than the heart as the site of intelligence. And it was there that Euclid devised geometry; Archimedes invented his mechanical wonders and, if legend has it right, ran naked through the streets yelling "Eureka" ("I've found it!") after discovering displacement theory in the bathtub; Hipparchus mapped the constellations and measured the brightness of the Sun. The heart of the library was its papyrus scrolls, now thought to number in the many hundreds, which stored Western knowledge. One of them was written by Aristarchus of Samos two millennia ago and said the planet on which he lived was one of several that circled the Sun. That scroll, the contents of which were described in Athens, was destroyed with all the others and almost everything else when the library was deliberately torched by prototypical terrorists. As a consequence, thirteen centuries passed before Kepler, Copernicus, and Galileo picked up the trail that led back to heliocentrism. "From the time of its creation in the third century B.C. until its destruction seven centuries later," Sagan lamented, "it was the brain and heart of the ancient world."

Nature has an infinite capacity for violence and destruction, but humanity has surpassed it where erasing civilization's record, and the ecosphere in which it exists, are concerned. Irreplaceable documents, including firsthand accounts of historically important events, have been lost to the destructive impulses of vandals and arsonists, the greed of desecraters and looters, and the grim requirements of war. So have unique, often priceless, artistic and literary masterpieces, musical instruments, and architecture. Medieval illuminated manuscripts have been cut up to be sold as wall art, and ancient artifacts from Afghanistan to Africa have found their way onto the international rare-art market. Archaeological sites throughout Iraq were plundered during and after two wars, with many important objects either destroyed or deliberately broken into pieces to be sold separately. Likewise, the National Museum of Iraq in Baghdad has been hit twice, with nearly 10,500 precious objects taken in and after the 2003 war.

The threat from war has escalated with the destructiveness of the weapons, cultural hatred or disdain, and the widespread poverty and frustration that turns the victims of oppression into vengeful scavengers, looters, and impulsive destroyers. The shelling of the Parthenon was thought to have military virtue. So did the bombing of Dresden and the eradication of most of Hiroshima and Nagasaki. Ultimately, there remains the prospect of a major nuclear conflagration; of scores or hundreds of radioactive infernos that would consume libraries, museums, galleries, and archives of every kind.

And terrorism has now been internationalized by large gangs of irregulars who glory in death and destruction, and so much the better if what is destroyed is a cultural icon of the enemy's. The leveling of the World Trade Center and the attendant loss of almost three thousand lives, hundreds of businesses and their records, and valuable art is one obvious case in point. And there is the pilfering and destruction by still others of their own culture in Iraq after two wars there in a dozen years. Archaeologists sifting through the rubble in the National Museum, and at "digs" around the country, were often in despair that brought tears as they surveyed sites that were destroyed by scavengers with crowbars and shovels. One archaeological site about three hours south of Baghdad had so many looters working at night that one observer likened it to a city of lights. Much of what was taken—delicate cuneiform tablets with their writing immaculately preserved—had not only not been studied, it had not been carefully recorded or photographed. And as one saddened archeologist said, an earthen cuneiform tablet driven over bumpy roads in the back of a pickup arrives at its destination in its original form: dust.

To all that, add normal decay and the dangerous caprices of an electronically digitized world. That pages in books progressively dry up, turn yellow, and start to disintegrate in less than a century makes the first point. The great North American blackout of mid-August 2003 makes the second. Digital technology's great strength is that it can accumulate, store in a small space, and electronically move colossal amounts of information around the globe. But that is also its great vulnerability, as everyone whose computer has ever "crashed"—become instantly dysfunctional and lost some or all of its memory—knows. And it can happen not only to individuals, but to businesses and other institutions, including cities and towns, entire regions, governments, and entire nations as well. Causes can range from electrical power blackouts like that one in Ohio on August 14, 2003, which quickly spread to the northeastern United States, southeastern Canada, and elsewhere, to massive lightning storms, to terrorists and others either taking down whole computer networks or infecting them with powerful viruses.

We are now so dependent on computers that a massive failure could result in what Bill Joy, the cofounder and chief scientist at Sun Microsystems, has called a dysutopian society. Technological chaos might come in the form of a massive breakdown in the globalized computational system, either accidental or deliberate, causing widespread disruption of communication, transportation, manufacturing, and other vital services that could in turn cause severe damage on a planetary scale. The damage would necessarily extend to electronically concentrated scientific and cultural information as well.

In that regard, and with supreme irony, the Library of Alexandria is back and it's digitizing. In fact, it's digitizing with a vengeance, as if to make up for having been "down" for more than fifteen hundred years. The reincarnation, a steel and glass structure with seven terraced floors for reading rooms that opened in October 2002 with a quarter of a million standard books, is planning nothing less than to make every book in the world available at the click of a mouse. Its directors have begun the Alexandria Library Scholars Collective, an unprecedented attempt to use new software to make all books accessible to students and scholars everywhere. That goal is splendid. But as is the case with other large caches of relics of Earth's history—the Library of Congress is converting 2.6 million phonograph records in all three speeds to the digital format, and the American Museum of Natural History is doing the same thing with its roughly 2 to 3 million fossil records, to take only two examples—the electronic format brings the potential for catastrophic loss as well as the advantage of quick and easy accessibility. The examples of information being saved, which include biological and botanical specimens, are nearly infinite.

The way to protect a civilization whose total information system is as vulnerable as ours is to take the precaution used by the people who operate the computers: back up the information and store it in several archives, both on and off the planet. Two underground repositories, one on either side of the globe, would hedge against major destruction in either hemisphere. A space station would be another likely place for an archive. And another would be in a large lunar colony.

The idea of having a comprehensive copy of what Earth is in its many forms, including its entire cultural spectrum and the natural environment in which it exists, was alluded to by Isaac Asimov's reference to an *Encyclopedia Galactica* in his Foundation series. And the bronze plaques carried by *Pioneers 10* and *11* that describe where they come from and the race that made and launched them into the cosmos are a rudimentary way of doing the same thing. Both spacecraft scouted Jupiter and Saturn in advance of *Voyagers 1* and *2*, which carried far more detailed discs, with Earth sights and sounds, and a diagram of the solar system from which they came. The fundamental difference between the data carried by the four spacecraft, aside from the quantity of information, and the system described here, is that the Pioneers and Voyagers are carrying data about Earth for other civilizations, not for use by this one.

The version of a comprehensive record discussed here was conceived by Professor Robert Shapiro, then a biochemist at New York University, in the late 1990s. Shapiro has an extraordinary imagination. He has pondered and written about life elsewhere and, in a book called *Planetary Dreams*, described the age-old fascination humans have had with the possibility of life existing elsewhere. Having gone through the Cold War and been aware of the threats to Earth, terrestrial and otherwise, he has also pondered the possibility of life ceasing to exist here, or else being caught in a calamity that inflicts widespread death and destruction. In such a situation, Shapiro decided, civilization's record could be lost for all time.

"I seldom walk away from my computer for any length of time leaving a morning's or an hour's work unsaved; some momentary flicker of the electrical system might wipe it out," he has noted. "My more ambitious efforts, full length books for example, are not simply left to the care of the hard drive, which can fail, but are always backed up. . . . I secure the value of my home by buying insurance, not in the expectation that it will burn down, but simply to protect myself if that unlikely possibility should come up. And vital documents are stored off premises in a more secure location, such as a safe deposit box in a bank. Yet collectively we plunge into the future with hardly a thought about the ultimate security against hazard of our civilization at full speed, with very limited visibility ahead, and no insurance at all.

"Common sense and historical precedent suggest one solution immediately," Shapiro reflected. "During the eclipse in Western culture following the fall of the Roman Empire [which some have theorized was caused by an impact], many works of the ancient world were preserved in Muslim societies or in sequestered monasteries, and later reintroduced into the West. In a similar manner, we need to have a sanctuary; a place of safety where a viable fragment of humanity can weather any coming storm and transmit the full record of our civilization to our descendants." That is how what soon came to be called the Alliance to Rescue Civilization, or ARC, was born.

Since a number of really nasty things that can happen to Earth would affect the whole planet, or most of it, ARC's purpose is to start an all-inclusive archive at locations scattered around Earth, and also in the lunar colony. And there are important caveats about the archive. The first, and by far the most important, is that it should, as part of planetary defense, constitute the single, focused purpose for returning to the Moon. The idea of going back in sizable numbers stalled after Apollo in large part because the competing reasons for doing so effectively canceled each other out; the space community's reasons for returning were so fragmented and competitive that they nullified each other. The competing reasons—commerce, science, adventure, and so forth—have been articulated for many years. But the net effect of so many individuals and groups with different agenda wanting to return prevented any momentum at all. The founding members of ARC—a half dozen individuals with varying backgrounds are working on it—are convinced there is no more compelling reason to return to the Moon than to save civilization. This is more than reason enough to do so, and if that strategy is made to ride piggyback on resource exploitation or something else, its overwhelming importance will be lost in the compromise. At the same time, while no previously advanced excuse for going back to the Moon seemed sufficiently compelling to create a powerful constituency, the planetary defense program would ultimately open the way for all of them.

And since a long-duration rescue effort would have to be mounted if a global or near-global catastrophe occurs, keeping the archive on the Moon makes more sense than siting it on Mars for two reasons. The Moon should and will almost undoubtedly be settled before Mars because it is three days away, not six months to a year. And that, in turn, means rescuers would take three days to get to Earth, not several months. To guard against the worst possibilities, the lunar colony would have to support enough human beings to start repopulating Earth. The archive would be used to restore civilization.

Furthermore, the archive should not be a time capsule like the large collections of documents and memorabilia people with spatulas and a little cement used to seal in the cornerstones of important buildings. That would defeat the archive's purpose, which is to back up the collective hard drive for use in saving and resurrecting a stricken region or most of the planet. A time capsule cemented in a building in 1900 would contain no information about the first two world wars, the Russian Revolution and the eventual collapse of the Soviet Union, the development of nuclear energy for peace and war, airplanes, the end of colonialism, Albert Einstein and relativity, Salk vaccine and the eradication of polio, all of the other medical, scientific, and engineering advances, the civil rights revolution, Pablo Picasso, Walt Disney, the World Trade Center and its destruction, landing on the Moon and the exploration of the solar system and beyond, the discovery of dark matter, Harry Potter, films, television, computers, nanotechnology, Nazism and the Holocaust, Babe Ruth, the Gershwins, the Beatles, hip-hop, and all the other things the twentieth century gave the world for good or ill. The new archive would therefore have to be updated continuously or at regular intervals as a living entity, not treated as an all-but-useless historical relic.

The archive concept is currently under study. Andrew Turner and a colleague are convinced that the internationalization and spread of terrorism poses a growing threat to corporations and to the commercial sector in general. In the time since the attack on the World Trade Center, the loss of transactional business and other data has closed stock exchanges and caused many business failures, he has noted, adding that it is crucial that important information be preserved even in the most dire emergencies. Turner, too, is therefore looking at the Moon as the likely site of an archive that businesses could use to assure the safety of their records. "Use of a cache on the Moon would enable network-free access of archived data in the event of a catastrophe which may very well shut down the Internet and other networks," he has explained. There are secure places in the United States, such as caves, but they could only be accessed from business centers in cities either by networks or by physically transporting discs. "These sites are not in direct line-of-sight view of New York, and if they were, they would be less secure since they would be vulnerable to future large-scale terrorist attacks that targeted major cities," Turner said. He also rules out airplanes for data storage, since they are dependent on an air traffic system that could be knocked out in an attack, and even satellites could be targeted by ground-based lasers or nuclear-tipped missiles. A payload installed on the Moon, on the other hand, could be accessed by line-of-sight communication from any place on Earth for about ten hours a day. In an emergency, portable antenne could be mounted and pointed to the lunar cache to pull in important data. And the lunar site itself would be hardened against any conceivable directed-energy attack from Earth. Says Turner, "The Moon can function as a new type of communications node never available before in the history of telecommunications. Unprecedented in its security and availability, the lunar data cache would provide new confidence for a troubled world." AT&T has announced that it is also trying to think up ways to protect business records, though they are rudimentary at this point.

Archiving would also provide a technological challenge because of changing methods of communication. There are probably reel-to-reel tapes, long-playing and other phonograph records, home movies, and some kinds of videotapes in the many millions that can no longer be played because there is no means to do so. The "miraculous" high-fidelity records of the 1950s are now almost entirely unusable antiques. Nanotechnology will accelerate that process. The solution where long-term, off-Earth archiving is concerned would seem to be the creation of an infinitely adaptable communication system that would change as technology advances, while either updating the existing cache to keep up with the changes or else continuously adjusting to be able to access the old stuff.

A lunar archive for cultural and corporate records, and a settlement large enough to make use of them and repopulate and reinvigorate Earth in the event of a catastrophic emergency, will necessarily take a great deal of time and treasure to establish. The archive itself would, by definition, have to operate indefinitely. And that creates an inherent problem. The archive's existence cannot be held hostage to the vagaries of partisan politics. There is no guarantee that any political system, let alone a government, can endure indefinitely. That being the case, the archive would have to be supported mostly by the private sector, probably in the form of a combined endowment from several foundations, as well as the kind of corporate gifts that are bequeathed to hospitals and universities. This would be in keeping with the private sector's increasing move to space.

Professor Shapiro has devised a way of convincing well-off individuals and foundations to contribute generously to the cause. "Would you rather have your name on a dormitory that will be demolished in fifty years," he would ask prospective donors, "or on the Moon for eternity?"

# : 8 :

# THE GUARDIANS

The manned space program is in shambles. Indeed, if a program, in this sense, is defined as a comprehensive undertaking with an articulated, coherent goal, there is no manned program. There hasn't been one since Apollo.

Nine days after *Columbia* broke up over Texas on its approach to the Kennedy Space Center, *The New York Times* ran a story by Todd S. Purdum, a reporter in its Washington Bureau, which stepped back from the breaking news and surveyed the bigger picture. He reported that the dream of space travel grips much of the world, in part because of the export of American culture, which celebrates it in films such as *Apollo 13* and *Armageddon*. Yet space travel is no longer celebrated in America.

It is popular in Europe and Asia. "But here in the United States, the current reality boils down to commemoration of past achievements or fanciful speculation about a future that has receded again into science fiction barely 40 years after human space flight began," Purdum observed accurately. "Space has no big constituency, NASA's budget has been flat for a decade, and no president has put space travel front and center on the national agenda since Lyndon B. Johnson. Whatever the cause of the *Columbia* disaster, experts in and out of government see little chance for the robust debate or investment that will allow the space program to do anything soon but muddle on." George W. Bush's announcement of a "space initiative" on January 14, 2004, at least got the subject back on the national agenda, though it was noticeably lacking in specifics. That led a number of observers of the space program to think it was more about feel-good politics in an election year than about serious purpose. More perniciously, it not only derailed the Orbital Space Plane, but a follow-on shuttle program, and other concepts to extend and make permanent the human presence in space.

John Noble Wilford, the *Times*'s veteran space reporter, made the same point as Purdum. With the Cold War over, and the space race with the Soviet Union won (whatever that meant), Wilford wrote, "The United States then turned its back on distant space as a destination for human exploration, and for the last 30 years not a soul has ventured more than 300 miles above Earth's surface." For young Americans, he continued, there are no longer soaring dreams of the final frontier.

Alan L. Bean, one of the insiders Purdum interviewed for his story, was an astronaut-artist who walked on the Moon on November 19, 1969, as a member of the Apollo 12 crew. He told Wilford the Apollo astronauts had taken it for granted that the program they started would continue with the construction of a lunar base and space stations as part of humanity's logical expansion to space for a permanent presence there. "At that time in our culture's history, we were doing the most that was possible to be done. We naïvely assumed that's what would continue, but it didn't," a disappointed Alan Bean reflected. "It's the normal thing for a culture, in history, that we respond to emergencies."

They are looming. The litany of dangers, from high-velocity boulders peppering the neighborhood, to resource depletion, to the spread of terrorism and weapons of mass destruction, to global warming and the multiple hazards it is causing, is growing. Yet we are caught in a dangerous predicament. Unlike the other creatures on this world, humans—at least some of them—have the intellectual capacity to understand the precariousness of the situation. But there is no capacity to respond to it with a long-term plan because, like the other creatures, humans are fundamentally—perhaps because of their evolution—incapable of projecting threats to the distant future and coming up with ways to reduce or avert them. It is the ultimate chess game, and we are playing it like wood-pushers. Ray Erikson calls this reflexive reaction to danger, with no long-range strategy for averting it, a "fight or flee" mentality.

E. O. Wilson, the naturalist, has his own theory, which he shared in a speech he gave to the members of the Foundation for the Future in August 2002 when they presented him with an award. "The human brain evidently evolved to commit itself emotionally only to a small piece of geography, a limiting band of kinsmen, and two or three generations into the future. We are innately inclined to ignore any distant possibility not yet requiring examination, however promising, or menacing." Wilson explained this in Darwinian terms: "For hundreds of millennia, those who worked for short term gains in a small circle of relatives and friends lived longer and left more offspring, even when—and this is the important part—their collective striving put their descendants at risk."

Humanity is now at potentially serious risk because of our collective inability to break the short-term-gain cycle. Danger—risk—is relative. We tend to believe that whatever time we live in is the most dangerous. That certainly was the case for those who lived through the carnage and mayhem of two world wars and the advent of nuclear weapons. If the prospect of the destruction of much or most of the world in a short time is the criterion by which a situation is called most dangerous, then surely the Cuban missile crisis in October 1962 would win that dubious distinction hands down. Had Kennedy and Khrushchev not behaved as prudently (and given that both had hawkish generals urging them to do otherwise, as bravely) as they did, a civilization-threatening conflagration of almost unimaginable proportion would have occurred. Both sides' nuclear missiles and strike aircraft were on hair-trigger alert at that frightening moment, as they were every day of the Cold War. And Near-Earth Objects are certainly a threat. But that has always been so. If anything, the relatively new awareness of the potential risk, and the ability to use space to reduce it, make the planet safer from a serious asteroid impact than ever.

But dangers, while relative, abound. Nuclear weapons still exist in the thousands, are proliferating, and are coveted by a steadily growing number of religious fanatics who worship death and who would not hesitate to use them or the other weapons of mass destruction to exterminate those they despise and go to heaven. To the fiends add the naturally occurring afflictions that have been described here, plus a two-headed technology that will increasingly have the capacity to enhance Earth and endanger it (Ray Kurzweil's and Martin Rees's voracious nanobot swarms perhaps being one of them, though that is unlikely). Whether all this will bring on some kind of worldwide catastrophe is not the issue. The issue is about increasing planetary protection to ensure that no matter what happens, gradually or suddenly, civilization has the means to evade annihilation and survive. The overwhelming majority of boats do not run into emergencies and sink, but no boat owner in his or her right mind would put to sea without a fire extinguisher, a lifeboat, and insurance. Similarly, most homes do not burn down, but those who live in them safeguard precious papers and other things in safe-deposit boxes.

That being the case, it is time for responsible individuals and groups everywhere to take Earth for the ship it is and begin a fundamental, evolutionary strategy to maximize its chances for survival. It is time to come up with a long-term plan because the multiple dangers and the means of salvation are coinciding. Here are some core points:

Using space to protect civilization and the planet on which it exists will not be achieved through popular acclamation. Most well-educated individuals, let alone the relatively uneducated majority, think of activities in space as an abstraction with no direct meaning for them. And, in any case, they have many more pressing concerns than saving civilization. Invoking Apollo as an example of what can be accomplished in space with public support is illusionary. The American people were generally supportive when their countrymen reached the Moon, mostly for nationalistic reasons having to do with pride in being citizens of the first society to alight on another world, and demonstrating its superiority over an enemy belief system. But that does not mean there was an inherent fascination with space. It is a mistake to confuse momentary jubilation with long-term resolve.

As the Columbia Accident Investigation Board made clear in its report, NASA has in the years since Apollo gradually suffered a systemic disintegration that accelerated after the end of the Cold War. This was partly the result of a lack of fundamental purpose within the space agency. While the board did not say so explicitly, the lack of purpose likely led to the "broken safety culture" that ended in the second tragedy. More important, the lack of purpose reflected an institutionalized indifference in successive presidencies and in a Congress that believed those of its constituents who were not actively involved in space, as was the aerospace industry, had many more pressing priorities. Professor John M. Logsdon, the longtime director of the Space Policy Institute at George Washington University and the accident investigation board's space historian, wrote in the report that the Cold War imperatives that had made NASA so potent a political symbol for John Kennedy and his immediate successors now seem gone for good. "No longer able to justify its projects with the kind of urgency that the superpower struggle had provided," Logsdon observed, "the agency could not obtain budget increases through the 1990's. Rather than adjust its ambitions to this new state of affairs, NASA continued to push an ambitious agenda of space science and exploration." That is because the space agency's management has always believed, correctly, that it was bringing prestige to the nation and, incorrectly, that ambitious goals would be rewarded with the means to achieve them.

The old debate between advocates of manned spaceflight and unmanned missions is now hackneyed and ought to be put to rest. The argument that using people in space is needlessly expensive is valid only if they are doing frivolous things. But not only is there nothing frivolous about using people to protect their civilization, there is no alternative. That being the case, humans and machines must be coordinated so each does what it is best suited to do, hopefully in support of the other. The Lunar Orbiters, Rangers, and Surveyors that reconnoitered the Moon so men could safely land on it, who in turn set up machines to collect scientific data, is a textbook example of a mutually supportive man-machine effort.

Given a public that is largely distracted by terrestrial problems (ironically, many, such as famine, energy shortages, and nasty disputes over water could be eased or ended by using space), and an agency that is in the doldrums because the political power structure doesn't believe it has a serious purpose, it is time for a fundamental change in attitude in the upper echelons of government. The White House and Congress must recognize that protecting Earth is an issue so profoundly important it is beyond partisan politics. Understanding the magnitude of what is at stake—as a growing number of scientists and others are doing—and acting on it by forging a long-term planetary protection program will mark the difference between the merely self-serving political drone and the real statesman or -woman. The process should start with the creation of a commission of highly knowledgeable experts, including sociologists and others who have already projected the likely effects of an environmental catastrophe, as well as scientists, economists, and ecologists. And the deliberations should be made by experts from around the world. The problem, by definition, cannot be addressed unilaterally. The citizens of Earth deserve as much.

Planetary defense should be conducted not as a major program within the space agency, but as the agency's highly focused, overarching mission. This is not to say space science and exploration should be abandoned or subverted to "applications." Learning about the world in its entirety is fundamentally important for spiritual reasons that usually have ways of becoming practical. Exploration, whether accomplished by sending robots and people to other worlds, or by observing them with telescopes, microscopes, and colliders—"atom smashers"—is so important that calling it off would contravene human nature. It is inconceivable. But the core mission, in its totality, would send humans and robots to space for mutually supportive operations specifically designed to protect the planet. That is to say, NASA, its collective foreign counterparts, and other cooperating U.S. agencies should assume the role of Earth's guardians. As it is, the space agency continues to try to ensure its survival by appealing to a broad range of interests. NASA's Web site lists its traditional missions as exploring the universe, searching for life, and "inspiring the next generation" (presumably to take to space). Lately, the site has added, "To understand and protect our home planet." That is a promising, if small, step in the right direction. The idea is not new, but it has consistently been ignored because of higher priorities, most of them political.

While planetary protection per se was not a concept that floated around Capitol Hill in the 1980s, it was reflected in the NASA Authorization Act of 1988, also known as the Space Settlement Act of 1988: "The Congress declares that the extension of human life beyond Earth's atmosphere for the purposes of advancing science, exploration, and development will enhance the general welfare on Earth and that such extension will eventually lead to the establishment of space settlements for the greater fulfillment of those purposes."

The legislation was submitted by the late Republican George E. Brown Jr. of California and was written by his aide for space policy, Steven M. Wolfe, an unabashed believer in humankind's future in space. Wolfe became so angry when he read an article in *The New York Times* in late January 1993 claiming that human spaceflight served only Cold War interests and no longer seemed relevant in its aftermath, he sent an angry rebuttal to the newspaper. "The idea of human space exploration did not begin with the Cold War. As early as the turn of the century, learned men seriously contemplated piloted space travel," Wolfe wrote. "Visionaries of today have dreams of space far grander than their predecessors, and their spirit is no less resolved to turn their visions into reality. Their dreams include returning people to the Moon—this time to stay; creating vast new industries off our planet's surface using only the resources found in space; and building colonies in space capable of supporting thousands, and eventually millions, of homesteaders. We dream these dreams not only in the interest of science and commerce, but because they are the only direction in which the human race can evolve."

The evolution has to involve machines and people working together to complement each other. Machines working on their own would monitor natural resources, both to inventory them and to spot the sort of greedy spoilers who are destroying the Amazon and other forests and overfishing around the world. Imagery showing a wilderness being looted would immediately be turned over to the law enforcement officials of that nation. They might or might not act on the evidence, but the Planetary Protection Program would have done its part. Similarly, in the likely event that limits are set by the international community on the amount of fish that can be taken in international waters until the decimated schools are replenished, imaging satellites would be used to find and track plundering fishermen.

Spaceguard, with its centralized reporting point at the Harvard-Smithsonian Astrophysics Center, and with cooperative arrangements with observatories in other countries, should be the model for NASA's Planetary Protection Program as a whole. And Spaceguard should receive enough funding so it can build and operate telescopes for use around the globe and in space that will allow it to pick up and catalogue potentially dangerous asteroids and comets. The importance of the program is finally beginning to take hold. Spaceguard was originally mandated to find and catalogue almost all asteroids and comets that are a kilometer or larger. But depending on velocity and composition, even a 140-meter rock could strike Earth with a force equivalent to tow of the larges thermonuclear warheads ever made. Such an impact would effectively obliterate New York, Paris, Moscow, or Beijing. But Spaceguard's annual budget averages roughly about $4 million (plus a little extra from other sources to operate the telescopes), which ought to embarass politicians who allocate many times that amount for dubious projects, one of which included an infamous pork barrel deal to build a bridge in the Alaskan wilderness that goes to an uninhabited island. Many times that amount was spent to repair the sliding doors on the Vehicle Assembly Building at the Kennedy Space Center. Locating and cataloguing objects in the 140-meter range will require substantially more telescopic and other collection capability than currently exists. And that, or course, will require a funding level to support such a system. In the event an Earth-crosser is calculated to be on a probable or definite collision course with Earth, it would have to be moved off course. That job could go to the U.S. Air Force, whose expertise in the arcane business of "taking out" enemy satellites with antisatellite weapons would make it a clear candidate for doing the same to "enemy" asteroids. The airmen should coordinate asteroid and comet defense with Spaceguard's astronomers, and therefore with NASA, to develop a defensive plan that would be an integral part of the larger Planetary Protection Program.

Civilians, including scientists, engineers, and at least two former and current astronauts not affiliated with Spaceguard or the air force, have been working on ways to spot and intercept potentially threatening asteroids and comets, some of them impressively imaginative. A team of seven aerospace specialists from NASA's Langley Research Center in Virginia and the private sector have come up with a Comet/Asteroid Protection System, or CAPS. They are worried that the conventional ground-based telescopes used by Spaceguard are not only inadequate for providing 100 percent coverage of the big nasties, but that they would give no warning time for asteroids smaller than a kilometer, or for the long-period and smaller short-period comets that come from the Oort Cloud and elsewhere that could cause serious regional destruction. They consider this to be a serious gap in Earth's defensive system and have proposed to remedy it by sending detectors to space to work in conjunction with the ground-based telescopes. The idea would be to expand the range of detectable objects by searching the whole celestial sky regularly with a small constellation of spacecraft for permanent, continuous asteroid and comet monitoring. That way, the orbital trajectories of threatening intruders could be altered relatively quickly to avoid an impact. "A space-based detection system, despite being more costly and complex than Earth-based initiatives, is the most promising way of expanding the range of detectable objects," they told an audience at a major space meeting in Houston in October 2002.

A historic step in awareness of the NEO threat and ways to deal with it took place at another conference that was held in Garden Grove, California, on February 23–26, 2004. The meeting, called The 2004 Planetary Defense Conference: Protecting Earth from Asteroids, was sponsored by the American Institute of Aeronautics and Astronautics and the Aerospace Corporation. It was attended by more than one hundred astronomers, other scientists, and interested individuals who were not scientists. More than eighty papers were presented, addressing such topics as "NEO Impact Scenarios"; "Radar Reconnaissance of Potentially Hazardous Asteroids and Comets"; "Space Impacts Mitigation: Deflection and Dispersion Based on Nuclear Explosions"; "The Mechanics of Moving Asteroids"; "Psychological Factors Influencing Responses to Major Near-Earth Object Impacts"; "Communicating the Unimaginable and the Effects of Pop Culture on Catastrophic Disaster Perception"; and the "B612 Mission Design."

The B612 Foundation, a nonprofit group whose mission is to develop and demonstrate a capability to deflect asteroids before impact, was founded in the autumn of 2002. It is named after the asteroid in Antoine de Saint-Exupéry's classic children's book, *The Little Prince*. Four of the founding members—Russell L. "Rusty" Schweickart, a former astronaut; Edward T. Lu, a current astronaut; Piet Hut, a scientist at the Institute for Advanced Studies in Princeton; and Clark Chapman—wrote a richly detailed and imaginative article for *Scientific American* that described the threat posed by asteroids and a way to physically reduce it. They came up with a plasma-fueled asteroid "tug" that would land on a potential impactor more than a decade before a probable collision and gently nudge it off course the way tugboats in harbors nudge ships.

They have suggested demonstrating the technique on an asteroid before 2015 and make the point that it would allow more control than using nuclear explosives, solar pressure from photons, and other methods. (Such a rendezvous has already happened. On February 12, 2001, the Near-Earth Asteroid Rendezvous spacecraft, which was on a 2-billion-mile science mission, landed on a twenty-one-mile-long asteroid named Eros for on-the-spot inspection after studying it in close formation for a year.) In common with other knowledgeable individuals who are considering the problem, including those in Spaceguard, the B612 group knows that, while not a threat to all of civilization, a hundred-meter asteroid impact could cause horrendous death and destruction to a densely populated region. And like everyone else who is studying the potential threat, they also know that what cannot be seen and plotted cannot be deflected. Locating and cataloging asteroids and comets, large and relatively small, therefore remains the first priority.

The National Reconnaissance Office looks for trouble in the other direction: down. If protecting civilization includes preventing superweapon proliferation, and the horrendous destruction and loss of life that can result from it, the NRO needs to continue to be heavily funded to carry out its space reconnaissance mission. And while there can be no suggestion of revealing classified information, especially since planetary defense by definition must be an international effort, the NRO, too, should continue to communicate and share imagery with its civilian counterparts, as it has done since the 1960s, and through the suitably ambiguous-sounding Civilian Applications Committee starting in 1975. NASA can continue to reciprocate, as it has during that time, when one of its imaging satellites spots something that is of interest to the intelligence community. Other intelligence organizations, including the Department of State's Bureau of Intelligence and Research, the Department of Energy's Office of Intelligence, and the Defense Intelligence Agency, should also participate. Coordination should benefit the civilian and military sectors and be synergistic. The key would be to share information for the best possible understanding of the threat. While "compartmented" information has long been used in the intelligence community to limit damage by a betrayer, it has also caused a number of serious errors and misunderstandings.

International terrorists, who had been solidifying for a decade as they bombed selected targets, officially raised their curtain on September 11, 2001. Two years later, with the U.S. invasion of Iraq as a pretext, deeply angry militant Muslims joined the ranks of the existing worldwide cadre of those who want to destroy the West at any cost. One obvious way to inflict terrible damage, both by killing innocent people and terrorizing others, is through weapons of mass destruction. As noted in the case of North Korea's nuclear weapons program, space reconnaissance has an important role to play in stopping superweapon proliferation. Chemical and biological weapons production are more difficult to find because, unlike nuclear weapons, they do not require relatively elaborate manufacturing facilities and can be made with the same kind of off-the-shelf hardware that is used for bacterial and virus research and pesticides and other chemical concoctions. (Though as previously noted, radically smaller equipment for making fissile material out of raw uranium, the Zippe centrifuge, was smuggled and sold to at least three nations by Pakistan's A. Q. Khan.) It is therefore unrealistic to believe that imaging satellites, no matter how capable, will ever be able to locate carefully hidden facilities.

Yet other spacecraft will have an increasingly important role to play in combating terrorism. These are the spacecraft that do not look, but that listen. But since no machine can read a terrorist's mind, Western intelligence agencies—in collusion with their Russian counterpart—have stepped up their infiltrating of terrorist organizations and tracking of their leaders. The number of international killers who have been caught and successfully interrogated since September 2001 bears testimony to the campaign's success. (The bad news is that they are replaced almost immediately by their lieutenants.) Part of that campaign has been and will continue to be eavesdropping on their communications from space. Large communication-intercept satellites continue to be launched into Clarke orbit from the Kennedy Space Center on their ultrasecret missions. What is needed is a new class of much smaller listeners that can be produced in quantity and deployed, much in the manner of the fleet of communication satellites, to pick up local conversations and other electronic interchanges between individual terrorists and their cells, and to triangulate their positions so they can quickly be found and caught. That will be one of the key missions for the secret robots until the threat from international terrorism is ended, if ever.

The core of the Planetary Protection Program is the lunar base, to be expanded into a growing colony, and to provide a home for the archive. President Bush was therefore right in calling for a return to the Moon in his space initiative address to NASA. Certainly it should be explored in the search for increasingly promising signs of life, and also just for the sake of getting to know it in greater detail. But people ought to be sent to the Moon not to use it as a staging place for a journey to Mars, but to turn it into a habitat. And Bush was seriously misadvised about using the Moon as an embarkation point for an expedition to the Red Planet. It is true that because the Moon's gravity is only a sixth of Earth's, it is cheaper (that is, it requires less velocity, hence less thrust) to leave it than to reach orbit from Earth. But when the cost of going from Earth to the Moon, using retrorockets to land on the lunar surface, and then burning still more fuel to get back off and headed for Mars or somewhere else is factored in, it would cost about twice as much to head for Mars from the Moon as it would to go directly from Earth. Saunders B. Kramer, who was intimately involved in a number of important classified and open space projects when he worked for what was then called the Lockheed Missiles and Space Company during the Cold War, easily made the calculations. He therefore called the idea of going from the Moon to Mars "pretty damned silly."

As much has been said about the *International Space Station* as well. Most everyone who believes in colonizing space thinks that large stations should be an integral part of it, and certainly a part of settling the Moon. The key is to borrow a page from Gerard O'Neill: build a large one, make it as self-contained as possible, and populate it with a tiny community. But there are two obstacles. Such a station would be so expensive, especially with the current and projected budget deficit, that it could not be funded. In addition, there is no grassroots constituency for a huge spaceship, since the overwhelming majority of people do not think it is necessary.

The *International Space Station* is a feeble compromise. It was originally touted as an advanced *Skylab* on which perfect ball bearings would be manufactured in zero gravity and, like the shuttles, where experiments would be performed that *could* (notice the qualifier) help prevent or cure horrible diseases. Not to leave anything to chance, its prime contractor, the Boeing Company, even ran newspaper ads that showed a little girl saying that the station kept her father employed. In the end, the "space science" came down almost entirely to endlessly repetitive physiological experiments on astronauts to understand how the human body reacts to long-duration spaceflight. Successive crews are not given flight or mission numbers. They are given "expedition" numbers. The bioastronautics research, as it is called, does investigations on subregional assessment of bone loss in the axial skeleton on long-term spaceflight; promoting sensorimotor response generalizability: a countermeasure to mitigate locomotor dysfunction after long-duration space flight; and advanced diagnostic ultrasound in microgravity, to take only three that are described in medspeak. But the irony is that the only long-duration mission is the station's endlessly flying in circles around Earth. Nothing else is in the works, nor will it be for a long time, Bush's grandiose "vision thing" notwithstanding. It is worth recalling that before he announced his initiative, he reduced the station's crew size by half and lopped off a science module, angering the *ISS*'s other partners.

And it gets worse. There is an adage in the aerospace world that goes like this: with a constituency, you don't need a mission; without a constituency, the mission doesn't matter. The space station does have a constituency. It is the international consortium of companies that have profited from building it and which will continue to do so with restored access by the shuttle. There are so many fat contracts and subcontracts that in the most honest of all possible worlds, the *ISS* would be called the Great Pork Barrel in the Sky. That's what gives it a domestic constituency as well as constituencies in Europe, Canada, Japan, and Russia. What is more, completion of the station will almost undoubtedly be its kiss of death, since the big profits are in construction, not in operation and maintenance. That means the station's real mission is to be built. Once that has happened, especially given that tenuous initiative (which, if nothing else, will generate a mound of lucrative study contracts), the station will suffer a slow, neglected end.

That does not have to be humankind's fate in space. There has to be a true, continuous presence there; a presence that has a compelling purpose. And the continuous presence will, in turn, require much less expensive launch systems than are now in use. Yet however relatively inexpensive the launch vehicles and spacecraft that carry people and cargo to the Moon, the inescapable fact is that going there will be expensive indeed. But there is no alternative except what could be the ultimate catastrophe.

The most daunting obstacle to a permanent program to use space for the protection of Earth is not financial or technical. It is political. It is of utmost importance that a bipartisan planetary-defense culture takes hold in the United States and around the world and accepts that space budgets must not only grow, but must be stable and protected over the infinitely long term, rather than be debated and redebated every year. Planetary defense must, in other words, become as normative as the military. No government would consider abandoning its armed forces. Protecting Earth, as its constituent nations are protected, should become permanently institutionalized and financed accordingly. There is a model. The navy budgets the operation of large vessels, such as aircraft carriers, for the expected life of the ship. It is inconceivable that a $4 billion supercarrier, which takes seven years to construct, would not have enough operating funds so it could fulfill its mission over its projected lifetime. As the carrier admirals do not have to scratch for funding to operate their ships every year, neither should the managers of the spacecraft fleet, the stations, and the lunar colony. Their funding must be as steady and dependable as the military's.

Planetary defense is by definition international. As astronomers in Spaceguard cooperate with their foreign counterparts, so should the world's space agencies where the defense of Earth is concerned. National interests started the space age, prevailed during the Cold War, and continue. China's orbiting a man, promising to orbit others, and announcing that it is aiming at the Moon were blatantly nationalistic. Every nation with a space program uses it, at least in part, for national security and political leverage.

Yet there are areas in which cooperation for the common good could be accomplished, most likely under the auspices of the United Nations. Protecting Earth is an obvious one. The UN has a number of organizations devoted to international cooperation off Earth, such as the Committee for the Peaceful Uses of Space, and specialized agencies such as the International Telecommunications Union and the World Meteorological Organization. In *Goals in Space: American Values and the Future of Technology*, which came out at the end of the Cold War, William Sims Bainbridge made the point that joint space projects between nations improve international cooperation. That is generally, but not always, true. The *Shuttle-Mir* mission from 1994 to 1998, in which astronauts and cosmonauts trained together for flight on the *International Space Station*, was a case in point. A series of life-threatening crises, including a fire, chemical leaks, power failures, and a collision with a supply craft, have been widely reported. Less well-known was the behind-the-scenes hubris that two proud space programs, one having started the space age and the other having sent its citizens to the Moon, brought to the joint enterprise in a stubborn test of wills that created strained feelings and seriously hurt it. Yet, ultimately, the series of near disastrous accidents on the aged station taught its Russian and American occupants an enduring lesson about the space environment. There was no possibility of not communicating with one another or working independently, let alone stalking off in a huff. They had to cooperate or die. At this stage, just reaching space is hard, and so is living there. But that will subtly change as the expansion continues and cooperation increases.

The Alliance to Rescue Civilization should also enhance international cooperation. *Civilization* is all-inclusive, and the *alliance* means that every nation, religion, ethnic group, political persuasion, and profession should be encouraged to participate by adding what it considers its most important attributes to the common archive. There should also be independently researched accounts of events, large and small, that have shaped Earth's history and will continue to do so. It is imperative that this whole civilization, which may be unique in all of space and time, survive.

"From my perspective, if space settlements materialize as they are currently envisioned, they will be less interesting as engineering triumphs than as human accomplishments that will shape the lives of future generations," Albert A. Harrison, a psychologist with a long interest in the habitation of space, has written. "Space settlements are intended to solve human problems. If they evolve as we hope, they will offer safe, provident, and wholesome physical environments; political and social reforms; and abundant opportunities for residents to flourish materially and psychologically. A strong humanitarian bias contributes to our vision of space settlements."

Jonathan Schell, who wrote *The Fate of the Earth*, a classic attack on nuclear weaponry, is in accord. He has written with grace that there is a philosophical "view" about civilization's need to endure that surpasses even the view from space. "It is the view of our children and grandchildren, and of all future generations of mankind, stretching ahead of us in time—a view, not just of one Earth, but of innumerable Earths in succession, standing out brightly against the endless darkness of space, of oblivion. The thought of cutting off life's flow, of amputating this future, is so shocking, so alien to nature, and so contradictory to life's impulse that we can scarcely entertain it before turning away in revulsion and disbelief."

Martin Rees, whose crystal ball is cluttered with all manner of ghastly perils, is nonetheless another devout believer in the sanctity of humanity and in the absolute necessity of its rescuing itself in space. He does not think colonies in space will sprout like flowers in a field or be a panacea for population and other earthly problems. But he sees them as a crucial hedge against doomsday. "Even a few pioneering groups, living independently of Earth, would offer a safeguard against the worst possible disaster—the foreclosure of intelligent life's future through the extinction of all humankind," which remains vulnerable so long as it stays confined and isolated here on Earth, he has written.

"Are we really helpless captives between an irrelevant past and an obscure future?" Bruce C. Murray, the deeply thoughtful Caltech geologist and future director of JPL asked, rhetorically, in 1975. "I think not." As Rees was to do years later, Murray went on to define humanity and make a characteristically eloquent case for its existence: "We are those who lived before us and those who will live afterward, sharing a cosmic lifetime. We were not created instantly at birth; we began to be in a momentous event more than three billion years ago when the first self-replicating molecules formed by chance. We will not entirely die so long as our thoughts, our imaginative creations, persist faintly within the consciousness of distant descendants and their cultures."

The continuation of life transcends death, not only physically, but spiritually. The protection of Earth and the creatures on it, connecting the majestic and likely unique accomplishments that started and nurtured civilization, with the unimaginably humanistic achievements of the distantly born, is supremely ennobling because it honors and dignifies the precious thing that is life. How unspeakably sad it would be if the cultural riches future generations could bring to the world in art, literature, science, politics, all manner of scholarship, and perhaps a philosophy that enables peace and mutual support to finally take hold were preempted with no hope of being realized. Using space to protect civilization, providing an environment in which it is able to collectively thrive and grow to its limitless potential, will transform humankind from its traditional role as the hapless victim of fate to one better able to control its destiny and fulfill its inherent, and perhaps unique, potential for greatness.

Further, the unborn deserve to fulfill their potential even for what is far less than great. Everyone who believes in the migration to space, for whatever reason, accepts that people will bring their baser instincts with them just as surely as they will bring the higher ones. There is no reason to suppose that evil, stupidity, and unenlightened self-interest will be left on the home planet. What transgresses the morality of any given moment is the sanctity of life itself and the overarching need to protect and enhance it. Being less than perfect, being tarnished, is infinitely better than not being at all.

Survival is therefore imperative. That is why the humans who inhabit this cradle of life in a vast, dark universe have been given the means to protect it for themselves and for those who will come after them.

# APPENDIX A

## An Open Letter to Congress on Near Earth Objects

July 8, 2003

### United States Senate

**Appropriations Committee**
Hon. Ted Stevens, Chair
Hon. Robert C. Byrd,
  Ranking Member
**Commerce, Science, & Trans. Com.**
Hon. John McCain, Chair
Hon. Ernest F. Hollings,
  Ranking Member
Hon. Sam Brownback, Subc. Chair
Hon. John B. Breaux, Subc.
  Rank. Mem.
**Armed Services Committee**
Hon. John W. Warner, Chair
Hon. Carl Levin, Ranking Member
**Foreign Relations Committee**
Hon. Richard G. Lugar, Chair
Hon. Joseph R. Biden, Jr.,
  Ranking Member
**Governmental Affairs Committee**
Hon Susan M. Collins, Chair
Hon. Joseph I. Lieberman,
  Ranking Member

### U. S. House of Representatives

**Appropriations Committee**
Hon. C. W. Bill Young, Chair
Hon. David R. Obey,
  Ranking Member
**Science Committee**
Hon. Sherwood L. Boehlert, Chair
Hon. Ralph M. Hall,
  Ranking Member
Hon. Dana Rohrabacher, Subc. Chair
Hon. Bart Gordon, Subc.
  Rank. Mem.
**Armed Services Committee**
Hon. Duncan Hunter, Chair
Hon. Ike Skelton, Ranking Member
**International Relations Committee**
Hon. Henry J. Hyde, Chair
Hon. Tom Lantos,
  Ranking Member
**Select Com. on Homeland Security**
Hon. Chris Cox, Chair
Hon. Jim Turner,
  Ranking Member

**Re: The Imperative to Address the Impact Threat from Near Earth Objects (NEOs)**

Dear Members of Congress:

We write to you today as concerned citizens, convinced that the time has come for our nation to address comprehensively the impact threat from asteroids and comets. A growing body of scientific evidence shows that some of these celestial bodies, also known as Near Earth Objects (NEOs), pose a potentially devastating threat of collision with Earth, capable of causing widespread destruction and loss of life. The largest such impacts can not only threaten the survival of our nation, but even that of civilization itself.

Although we are genuinely concerned about the NEO threat, none of us is an alarmist. We know of no Near Earth Object currently on a collision course with Earth, but science's limited knowledge of the NEO population cannot rule out that possibility. Based on current information, a *crisis* response to these potential threats is not warranted. That being said, however, based upon evidence of past impacts and recent asteroid observations as well as the possible consequences from just one relatively "small" NEO impact, "business as usual" regarding this threat is simply no longer a responsible or sensible course of action.

Studies indicate that, with the commitment of modest resources, NEO impacts can likely be predicted and, with adequate warning, steps taken to prevent them. Thanks to scientific advances and increased awareness, we now have a historic opportunity to deal comprehensively and effectively with the NEO threat. Doing so, however, will require determined and coordinated action by Congress, the Executive Branch, and the private sector to direct effective use of our nation's substantial scientific and technological capability.

U.S. and international academic conferences, as well as Congressional hearings, have served to illuminate some aspects of the NEO impact hazard. Here, we build upon this background and outline a recommended course of action for Congress.

To address this potential threat, we strongly urge that each of you take steps within your respective committee jurisdictions to implement immediately the following recommendations (each is discussed in more detail in the enclosure):

1. **NEO Detection:** Expand and enhance this nation's capability to detect and to determine the orbits and physical characteristics of NEOs.

2. **NEO Exploration:** Expand robotic exploration of asteroids and Earth-approaching comets. Obtain crucial follow up information on NEOs (required to develop an effective deflection capability) by directing that U.S. astronauts again leave low-Earth orbit . . . this time to protect life on Earth.

3. **NEO Contingency and Response Planning:** Initiate comprehensive contingency and response planning for deflecting any NEO found to pose a potential threat to Earth. In parallel, plan to meet the disaster relief needs created by an impending or actual NEO impact. U.S. government/private sector planning should invite international cooperation in addressing the problems of NEO detection, potential hazards and actual impacts.

## Overview of Confirmed NEO Impacts and Recently Detected NEOs

**Sixty-five million years ago,** a trillion-ton comet or asteroid only about six miles across struck what is now Chicxulub on Mexico's Yucatan Peninsula. That impact resulted in the extinction of at least 75% of Earth's species, including the dinosaurs.

**Thirty-five million years ago,** a comet or asteroid only approximately 3 miles in diameter struck Earth in Chesapeake Bay, about 120 miles southeast of Washington, D.C. That impact created a crater some 50 miles wide, changed the courses of many modern rivers and caused changes in groundwater aquifers that are still evident today.

**Fifty thousand years ago,** an asteroid just 150 feet in diameter, weighing approximately 300,000 tons, and traveling at 40,000 miles per hour struck Earth in what is today Arizona. Today, the crater from that impact, even after weathering, is still nearly a mile wide and 570 feet deep.

**About a hundred years ago, on June 30, 1908,** an object from space appeared in the morning sky over western China. It plunged through the atmosphere, glowing at a temperature of over 5,000 degrees F. Streaking over central Russia, the object's passage produced a deafening roar, preceded by a supersonic blast wave that leveled trees and houses in its path. As reported in the newspaper *Sibir,* this impact occurred "early in the ninth hour of the morning." Near the Stony Tunguska River, the object exploded in mid-air with an energy greater than a 10-megaton nuclear blast. The explosion devastated a region some 40 miles across, two-thirds the size of Rhode Island. Only a few people were killed in this sparsely populated region, but the story would have been very different if the object had hit a few hours later over Europe instead of the Siberian forest. The death toll in major cities such as St. Petersburg, Helsinki, Stockholm or Oslo might have reached 500,000.

**In 1947,** also in Russia, in the Sikhote-Alin Mountains, northeast of Vladivostok, a small meteor traveling at 31,000 miles per hour struck Earth's atmosphere, creating a fireball witnesses said was brighter than the sun. One of the fragments left an impact crater 85 feet across and 20 feet deep.

**In 1994,** the world witnessed the devastating effects that a large NEO impact could inflict on Earth. Astronomers who had observed the breakup of comet Shoemaker-Levy 9 then tracked its headlong crash into Jupiter, where it generated an explosion with an energy equivalent to a billion megatons of TNT. The resulting dust cloud in Jupiter's atmosphere swelled larger than our own Earth: a similar impact here would have destroyed our civilization and devastated life on this planet. Shoemaker-Levy 9 *was discovered just sixteen months before it hit Jupiter,* and its spectacular demise was a shot across our bow—a reminder that comets also can strike Earth. Comets, though less frequent visitors to Earth's vicinity than asteroids, strike with much greater kinetic energy, and comprise a small but significant part of the impact threat to Earth.

**On January 7, 2002,** the asteroid 2001 YB5 missed our planet by a little more than twice the distance to the Moon. If this 300-yard-wide, stadium-sized object, discovered *only 12 days before* its closest approach, had hit the Earth's continental landmasses, it would have destroyed nearly everyone and everything in an area about the size of New England. An ocean impact would also have spawned huge tsunamis, with the potential for damage to coastal areas beyond anything in historical experience.

The modest search efforts sponsored by the National Aeronautics and Space Administration (NASA) and the Department of Defense have detected a steady stream of close encounters. **On June 14, 2002,** asteroid 2002 MN, an object about 100 yards in diameter, passed within just 75,000 miles of Earth at a speed of over 23,000 miles per hour. 2002 MN was detected by astronomers at the Lincoln Near Earth Asteroid Research (LINEAR) search facility in New Mexico *three days after* its closest approach to Earth. Had this object struck Earth, it would have exploded with energy about equal to that of the 1908 Siberian impact near Tunguska.

**On July 5, 2002,** the LINEAR astronomers discovered another object, designated 2002 NT7, estimated to be over a mile in diameter. And in **November 2002,** astronomers discovered 2002 VU94, an NEO estimated to be over two miles across. While both objects pose no danger to Earth in the coming centuries, their recent discovery and large size emphasize the fact that many large NEOs *remain undiscovered.*

Scientists have realized for some time that Earth travels amid a "sea" of similar objects, large and small. NASA stated last year in Congressional testimony that we have detected only a little more than half of all NEOs larger than a kilometer in diameter. Prudence dictates that more be done to identify NEOs, and to obtain the scientific information necessary to divert any sizable NEO found to be on a collision course with Earth.

## The NEO Threat

The latest NEO close approaches are typical of the two dozen such encounters known to have occurred in the 20th Century. These are only a small fraction of the actual number that have occurred; most have gone completely *undetected.* Such approaches are commonplace in our part of the solar system. The late planetary geologist Eugene Shoemaker put it succinctly: "Earth exists in an asteroid swarm."

We know that since 1937, at least 22 asteroids have approached Earth more closely than did 2001 YB5, which missed by just twice the distance to the Moon. Five of those objects were larger than 100 yards in diameter. According to NASA, there may be as many as 100,000 NEOs with diameters of 100 yards or larger. Of those asteroids larger than 150 yards in diameter, about 250 are today *estimated* to be potentially hazardous. The United States has very *limited capability* to detect these smaller NEOs, which can nevertheless inflict substantial damage upon striking Earth. There is a significant probability (20%) of such an object colliding with the Earth during the next century.

Although the annual probability of a large NEO impact on Earth is relatively small, the results of such a collision would be catastrophic. The physics of Earth's surface and atmosphere impose natural upper limits on the destructive capacity of natural disasters, such as earthquakes, landslides, and storms: By contrast, the energy released by an NEO impact is limited only by the object's mass and velocity. Given our understanding of the devastating consequences to our planet and its people from such an event (as well as the smaller-scale but still-damaging effects from smaller NEO impacts), our nation should act comprehensively and aggressively to address this threat. America's efforts to predict, and then to avoid or mitigate such a threat, should be at least commensurate with our national efforts to deal with more familiar terrestrial hazards.

If space research has taught us anything, it is the certainty that an asteroid or comet **will** hit Earth again. Impacts are common events in Earth's history: scientists have found more than 150 large impact craters on our planet's surface. Were it not for Earth's oceans and geological forces such as erosion and plate tectonics, the planet's impact scars would be as plain as those visible on the Moon.

## Potential Misinterpretation of NEO Impacts

Even small NEO impacts in the atmosphere, on the surface, or at sea create explosions that could exacerbate existing political tensions and escalate into major international confrontations. For example, an atmospheric impact in 2002 produced a large, highly visible burst of light in the sky during the height of war tensions between nuclear-armed countries India and Pakistan. That high-altitude explosion happened to occur over the Mediterranean, just a few thousand miles from their disputed border region. Had that NEO impact occurred less than three hours earlier, it would have detonated over southern Asia, where its misinterpretation as a surprise attack could have triggered a deadly nuclear exchange. With military and diplomatic tensions at their peak in other areas of conflict in the world, the potential for a mistake is even greater today.

## Conclusion

For the first time in human history, we have the potential to protect ourselves from a catastrophe of truly cosmic proportions. All of us remember vividly the effect on our nation of terrorist strikes using subsonic aircraft turned into flying bombs: thousands of our citizens dead, and our economy badly shaken. Consider the ramifications of an impact from a relatively small NEO: more than a million times more massive than an aircraft, and traveling at more than thirty times the speed of sound. If such an object were to strike a city like New York, millions would die. In addition to the staggering loss of life, the effects on the national and global economy would be devastating. Recovery would take decades.

We cannot rely on statistics alone to protect us from catastrophe; such a strategy is like refusing to buy fire insurance because blazes are infrequent. Our country simply cannot afford to wait for the first modern occurrence of a devastating NEO impact before taking steps to adequately address this threat. We may not have the luxury of a second chance, for time is not necessarily on our side. **If we do not act now, and we subsequently learn too late of an impending collision against which we cannot defend, it will not matter who should have moved to prevent the catastrophe . . . only that they failed to do so when they had the opportunity to prevent it.**

Our nation, our families, and others around the globe deserve our best efforts to protect against the NEO impact threat. We urge the Congress to call on this nation's ready supply of talents and energies to responsibly address this threat. Our international partners also should be called upon to help meet this challenge, but the United States has a compelling responsibility to lead the way. Preventing an NEO impact is a vital mission for our nation's space program and for the American people. For the first time since Apollo, our astronauts should once again leave low-Earth orbit and journey into deep space, this time to protect life on our home planet.

We strongly recommend your prompt attention and *action* to address this too-long-ignored threat to the security of America and to the world. The accompanying recommendations are prudent and concrete steps each of you can now take to safeguard our nation. Your timely and effective response can protect the people of the United States and the world from the real threat posed by Near Earth Objects.

Sincerely,

Dr. Harrison H. Schmitt
Former Astronaut, U.S. Senator
Planetary Geologist

Dr. Carolyn S. Shoemaker
Lowell Observatory

David H. Levy
Jarnac Observatory, Inc.

Dr. John Lewis
Professor of Planetary Sciences
Planetary Geologist

Dr. Neil D. Tyson
Director, Hayden Planetarium
New York, New York

Dr. Freeman Dyson
Professor Emeritus, Institute
for Advanced Studies
Princeton University

Dr. Richard P. Hallion
Aerospace Historian

Dr. Thomas D. Jones
Former Space Shuttle
Astronaut
Missions STO-59, 68,
80 & 98

Bruce Joel Rubin
Screenwriter

Dr. Lucy Ann McFadden
Planetary Scientist
University of Maryland

Erik C. Jones
Amateur Astronomer

Marc Schlather
President, ProSpace

William E. Burrows
Author, NYU Professor
and Space Historian

Enclosure: NEO Detection, Impact Prevention and Mitigation Recommendations
cc: President George W. Bush
Vice President Richard B. Cheney
Hon. Kofi Annan, Secretary General, The United Nations
Hon. Colin L. Powell, Secretary of State
Hon. Donald H. Rumsfeld, Secretary of Defense
Hon. Tom Ridge, Secretary of Homeland Security
Hon. Sean O'Keefe, Administrator, NASA
**United States Senate:**
Jim Morhard, Staff Director, **Appropriations Committee**
Terrence Sauvain, Minority Staff Director
Jeanne Bumpus, Staff Director, **Commerce, Science and Trans. Committee**
Kevin Kayes, Minority Staff Director
Floyd DesChamps, Maj. Sr. Staff Mem. **Science, Technology and Space
    Subcommittee**
Jean Toal Eisen, Minority Staff Director
Judy Ansley, Staff Director, **Armed Services Committee**
Richard DeBobes, Minority Staff Director
Ken Myers, Staff Director, **Foreign Relations Committee**
Antony Blinken, Minority Staff Director
Michael Bopp, Staff Director, **Governmental Affairs Committee**

Joyce Rechtschaffen, Minority Staff Director
**U. S. House of Representatives:**
James Dyer, Staff Director, **Appropriations Committee**
Scott Lilly, Minority Staff Director
David Goldston, Chief of Staff, **Science Committee**
Robert Palmer, Minority Staff Director
Bill Adkins, Staff Director, **Space and Aeronautics Subcommittee**
Richard Obermann, Minority Professional Staff Member
Robert Rangel, Staff Director, **Armed Services Committee**
Jim Schweiter, Minority Staff Director
Tom Mooney, **International Relations Committee**
Robert King, Minority Staff Director
Uttam Dhillon, Staff Director, **Select Committee on Homeland Security**
Steve Cash, Minority Staff Director
National Academy of Sciences
National Academy of Engineering
American Institute of Aeronautics and Astronautics
American Astronomical Society (Division for Planetary Sciences)
Association of Space Explorers
American Geophysical Union
Defense Advanced Research Projects Agency
Minor Planet Center
National Space Science and Technology Institute
The B612 Foundation
The National Space Society
The Planetary Society
Space Foundation
Space Frontier Foundation
Smithsonian Institution

**Point of Contact:** To respond to this letter, request additional information about it or contact signatories, please refer to www.CongressNEOaction.org or call or fax Dr. T. D. Jones at tel. no. (703) 242-9256 or fax no. (703) 242-8935

## NEO Detection, Impact Prevention and Mitigation Recommendations

After assessing the nature and scope of the NEO impact threat, and in consultation with many leading authorities on space issues, we are recommending three steps to deal with the problem. The first will increase our nation's ability to detect an NEO impact in time, the second will lay the groundwork for deflecting such an object, and the third will help mitigate the consequences of an actual NEO impact. Such steps would entail a relatively modest commitment of resources, an investment warranted by the potential consequences of misjudging the NEO impact threat to Earth.

### Recommendation #1:
### Immediately Increase the Scope of and Funding for NEO Detection

The United States is currently engaged in a search for all NEOs greater than 0.62 miles (a kilometer) in diameter. The effort is producing results, but only a few dozen researchers are funded to conduct this basic survey. Resources committed to this work have been very modest and not commensurate with the potential threat; thus, additional investment in search programs is both appropriate and prudent. A dramatic improvement in the rate at which asteroids and comets are discovered would likely result if the United States were to increase the current level of funding, now at about $3.5 million per year, to at least $20 million annually.

We recommend that Congress take the following measures to enhance the search for NEOs:

· **Increase search activities for detection of NEOs 0.62 miles (1 kilometer) in diameter and larger.** Researchers estimate that only one-half of such NEOs have been located. The pace of identification should be accelerated. Support for Southern Hemisphere search activities may further increase the discovery rate and should be expanded. Even when NASA achieves its current goal of identifying 90% of large NEOs, the undiscovered remainder will, of course, still pose a potential hazard. Congress should direct NASA to pursue the search for all such objects to statistical completion.

· **Expand the search effort to include detection and tracking of NEOs smaller than 0.62 miles (1 kilometer).** NEOs such as 2002 MN (about a hundred yards across) are not currently the target of any formal search program. Rather, they are discovered as by-products of the search for larger objects. Because an impact of even a relatively small NEO could still destroy a major city, the United States should establish the goal of predicting any close approach to Earth by any asteroid larger than 200 yards in diameter.

· **Increase funding for the Minor Planet Center (MPC) to $1 million annually.** The MPC is responsible for the collection, computation and dissemination of the characteristics and orbits of asteroids and comets. As the central international clearinghouse for tracking NEOs, it should be funded at a level more commensurate with its important role in understanding and addressing the NEO threat.

· **Provide funding for more and better instrumentation and additional follow-up observations.** In addition to maintaining existing optical and radar search programs, NASA should be given the added resources and mandate to enhance the instrumentation dedicated to NEO detection and to respond to NEO discoveries with more detailed observations. Such radar and spectroscopic observations are vital to refine asteroid orbits and determine an NEO's general composition.

## Recommendation #2:
## Expand Current NEO Exploration Programs

Given the real probability of an asteroid or comet impact, our nation must understand NEO characteristics well enough to develop practical methods to deflect them. Without adequate knowledge of the composition and mechanical properties of such objects, developing diversion strategies will be problematic at best and fatally ineffective at worst.

Therefore, we recommend that the United States take the following action:

· **Mount additional near-term robotic missions to selected asteroids and Earth-approaching comets.** By visiting NEOs in our own "neighborhood," we can determine their composition, measure their structural and mechanical properties, and provide the knowledge essential to preventing impacts on Earth by similar objects.

· **Begin planning now to send explorers to nearby asteroids and Earth-approaching comets.** Developing the capability to send astronauts to NEOs (on round-trips lasting just a few months) is the next logical human spaceflight goal for the United States. Such expeditions will help provide protection to Earth, serve as an insurance policy against future NEO impacts and, in the process, expand our ability to understand and use the vast and beneficial resources of space. Ideally, these voyages should immediately follow the completion of the International Space Station; planning for them should start now.

**Our nation should once again send its astronauts beyond low-Earth orbit . . . this time to protect our planetary home.**

## Recommendation #3:
## Develop NEO Contingency and Response Plans

Just as the federal government plans appropriate responses to disasters such as hurricanes and earthquakes, it should prepare contingency plans for dealing with an NEO impact. The government should begin planning now to deflect any NEO found to pose a potential threat to Earth. It should also plan to meet emergency response and disaster relief needs created by an impending or actual NEO impact. This government/private sector planning should include international coordination to address the issues of NEO detection, potential hazards and actual impacts.

To guide essential contingency planning, we recommend the following:

· **Establish an Interagency NEO Task Force to address the NEO Impact Threat:** This Task Force should be composed of senior representatives from appropriate government agencies: Department of Homeland Security; Department of Defense; Department of State; Department of Energy; NASA; Federal Emergency Management Agency; National Science Foundation; Office of Science and Technology Policy; and the National Research Council. The Task Force should also include appropriate representatives from industry and academia. It should be assigned responsibilities for guiding NEO impact contingency planning through an NEO Impact Response Center (see below), including identification, monitoring and analysis, international coordination of NEO search efforts, impact response and mitigation, and deflection strategies and technology.

· **Establish an NEO Impact Response Center:** This Center should be assigned responsibilities to—(1) collate accurate information from all available sources on the threat potential of any potentially hazardous NEOs; (2) distribute such information and analysis to public agencies, both in the United States and overseas; (3) develop and implement contingency plans, to include the actions required to deflect an NEO if that becomes necessary; and (4) ensure that an unexpected impact is not misinterpreted as an attack on any country.

· The Center should collect astronomical and technical data about NEOs provided by existing research and search efforts. More importantly, it should verify this information and provide authoritative analysis to the President (and Secretary of Homeland Security), and the relevant committees of the Congress in the event of a projected NEO impact. The Center would enable U.S. civil and military authorities to develop the appropriate responses to an impact prediction and disseminate impact information worldwide.

# APPENDIX B

## White Paper Summarizing Findings and Recommendations from the 2004 Planetary Defense Conference: Protecting Earth from Asteroids

## 1. Overview

The Planetary Defense Conference: Protecting Earth from Asteroids was held from February 23 to 26, 2004, at the Hyatt Regency Orange County, Garden Grove, California. The meeting was sponsored by the American Institute of Aeronautics and Astronautics (AIAA) and The Aerospace Corporation, and attended by over 140 participants (see list at end of document).

The conference was held to focus on mitigating the threat to humankind posed by asteroids and comets. It is well known that there have been impacts of large Near Earth Objects (NEOs) in the past history of Earth. It is also well known that, while the probability of impact in the next month or year is small, impacts of objects large enough to seriously modify or potentially end life, as we know it, are inevitable. The conference, expected to be repeated on a periodic basis, focused on what humankind might do to deflect an approaching object.

While perhaps the most broadly focused, the 2004 conference was not the first on the topic of NEOs and the threat they pose. Conclusions and recommendations from other conferences are generally consistent with those presented here. The fact that several recommendations are repeated demonstrates that progress has been slow in this area.

The February 2004 conference featured a systems approach to NEO deflection, risk communication, and disaster response. Experts in detection of NEOs, in possible methods of deflecting a threatening NEO, in mission design, and in political, policy, law, and disaster preparedness came together to assess the current state of knowledge in each of these areas relative to mounting a successful deflection mission.

At the conclusion of the meeting, session chairs provided summaries of principal recommendations from their sessions, and participants were invited to provide their thoughts. The conference chair, session chairs, and speakers subsequently prepared this document to summarize what they believe are consensus findings and recommendations from the conference. Names and affiliations of the primary participants in the development of this document are provided. This document reflects the expert views of this group.

## 2. Findings and Recommendations

Findings and related recommendations are provided in the five major topic areas discussed at the conference:

### A. Threat Detection and Characterization

Ongoing efforts to track and understand the nature of asteroids and comets were discussed at the beginning of the conference. Information on the size, composition, and dynamics of NEOs is critical to initially assessing and finally mitigating threats posed by these objects.

### Findings:

The risk of impacts upon Earth, chiefly by Earth-crossing asteroids, is small but very real. The potential damage depends on the impactor's mass, composition, structure (e.g., whether it is a single object; approximately 1/6 of Earth-crossing asteroids as large as a few hundred meters are thought to be binary bodies), impact speed, and the location of the impact point (inhabited land, uninhabited land, or the oceans).

Impacts span a huge range in severity and frequency, and the means to predict and mitigate these impacts vary accordingly. The probability of a "dinosaur-killer" impact is about 1 in one million this century. The probability of a civilization-ending impact is larger—a bit less than 1 in 1000 this century. The probability of a small or Tunguska-class impact (near the lower size for penetration of the atmosphere, but still large enough to destroy a city) is higher still: There is approximately 1 chance in 10 of such an impact this century.

The economic and social costs of an impact are difficult to predict, but are obviously important considerations in assessing the overall threat and justifying the cost of detection and mitigation measures. Costs for funding a fully adequate search program to detect and catalog threatening objects in the 100 m class and larger and funding research to develop deflection methodologies would total perhaps $1 per year per capita. Losses caused by impact of a NEO could be incalculable.

Current search programs are of a scale that can find Earth-crossing asteroids greater than one km in diameter—e.g., those whose impacts could destroy civilization. The U.S. Government's *Spaceguard Survey*[1] plans to detect and catalog 90% of NEOs one km and larger in size by the end of 2008. A search program to locate sub-kilometer Earth-crossing asteroids has been recommended by the National Research Council (which would focus on asteroids down to 300 m in size)[2] and by an internal NASA NEO Science Definition Team (which would focus on those down to 140 m).[3]

Comets are much harder to deflect due to their long-period orbits, short warning times, high velocities, and the non-gravitational forces on them—venting as comets approach the sun causes small, random deflections of their orbits, making accurate hazard predictions impossible and creating uncertainties in the nature of comets. Fortunately, the hazard from comets appears to be a very small fraction of the overall risk. Future conferences should continue to address the difficult but important task of detecting and mitigating comets.

Adequate warning time is a requirement to enable design and implementation of a mission to deflect a threatening object, and accurate detection and tracking are essential. Uncertainty in the risk of Earth impact decreases the likelihood that a mitigation mission could be proposed, funded, and successfully completed.

NASA currently has the responsibility to find and track NEOs. But there is no formal process in place for forwarding notice of a high-priority threat to agencies responsible for civil defense. Nor do civil defense agencies have plans for re-

---

[1] D. Morrison, ed., *The Spaceguard Survey: Report of the NASA International Near-Earth-Object Detection Workshop*, NASA publication (1992).

[2] "New Frontiers in the Solar System: An Integrated Exploration Strategy," Solar System Exploration Survey, Space Studies Board, Division on Engineering and Physical Sciences, National Research Council, 2003 (see http://books.nap.edu/html/newfrontiers/0309084954.pdf).

[3] "Study to Determine the Feasibility of Extending the Search for Near-Earth Objects to Smaller Limiting Diameters," prepared by Near-Earth Object Science Definition Team, August 2003 (see (http://neo.jpl.nasa .gov/neo/neoreport030825.pdf).

sponding to such notices. In addition, no national or international organizations or agencies are responsible for the more general problem of protecting the planet from impacts. Most organizations likely to be involved if a threat is detected are not even aware of this natural hazard.

## Recommendations:

1. Review current international amateur and professional efforts related to detection, timely sharing of survey data, astrometric followup, orbit calculation, and physical characterization of potentially threatening asteroids and comets, and develop recommendations for improving coordination of, and support for, these NEO activities.

2. Efficiently survey and catalog 100-m-class NEOs. The central conclusion of the NASA Science Definition Team report on NEOs is that the global residual hazard (that which will remain after completion of the current *Spaceguard Survey*) is reducible by relatively inexpensive telescopic and/or spacecraft systems. Such systems can rapidly retire most of the residual hazard for a fraction of the hazard's fiscal costs. However, a substantial increase in the funding base beyond the current level of NASA funding (~$4.0 million per year) is required to accomplish this survey of sub-kilometer asteroids, and this funding must be maintained into the future to watch for long-period comets and rogue asteroids.

3. Encourage the development of creative ideas for detecting and cataloging potentially threatening long-period comets.

4. Develop and fund ground-based techniques (including planetary radar) as well as missions to several asteroids to gather information that contributes to designing deflection missions. Critical information includes object sizes and dynamics, object types (e.g., binary), characteristics of surface and sub-surface materials, responses to explosive forces, and characteristics relating to attaching a spacecraft or other large structures to NEOs. This information is important not only for mitigating NEOs but also for enriching our scientific knowledge of asteroid properties.

5. Establish a formal protocol for disseminating information regarding NEOs when the probability of impacting Earth exceeds specified thresholds.

## B. Mitigation Options and Mission Designs

Several options for deflecting a threatening NEO were presented at the conference. These options are in various states of maturity—some might be available within a few years, others might require decades of development and testing before they could be used to move an approaching object. In some cases, technology developed for other space missions might be applied to moving an asteroid. Proposed approaches included using nuclear explosions or mirrors to ablate small amounts of material from the asteroid's surface, deflecting the object with a high-speed impactor, and attaching to the asteroid a high-efficiency electric propulsion system (whose development would be beneficial for a variety of deep space missions). Participants generally agreed that we need a test program to confirm our ability to move threatening asteroids in a controlled fashion.

## Findings:

The feasibility and method used to deflect an oncoming object depend on its size, spin state, and composition (e.g., solid body, loose aggregation of smaller bodies, binary body), as well as the time available to effect a change in the object's orbit. The range of all of these parameters is very great—the deflection means must be flexible enough to adapt to the particular NEO that threatens.

Nuclear explosives and kinetic impactors (for small NEOs) are possible options in the short term, though the effectiveness of either technique is uncertain, given the lack of appropriate testing for a mission of this type. The response of a NEO to a nearby nuclear explosion or a high-speed impact might be better understood from testing and with better information on the nature and composition of NEOs. (Current international agreements forbid testing or use of nuclear explosives in space, even for peaceful purposes. The effect of nuclear explosions could potentially be tested using non-nuclear means.) Use of nuclear devices would face significant political hurdles that could delay or derail a deflection effort.

Other deflection options, such as electric propulsion and laser ablation, have been proposed, but at current spending levels, many years would be required to develop these technologies to the point where they could be used for a NEO deflection mission. If we wish to avoid the launch of nuclear explosives into space, the development of these alternative technologies for NEO deflection should have increased priority.

The mission requirement for a NEO deflection effort (e.g., "reduce the probability of Earth impact to less than 1 in a million") will drive the design of the deflection mission. The possibilities of launch failure, sensor failure, and off-nominal performance by satellite and deflection systems, as well as uncertainties in the properties of the NEO, must be recognized and factored into the overall design of the mission.

To date, we have not demonstrated that we can actually deflect a NEO in a controlled way using any of the suggested mitigation techniques.

## Recommendations:

6. Conduct mission design studies to characterize requirements for short-, medium-, and long-range missions. These studies would compare current capabilities with mission requirements and help to identify and prioritize technology research and development goals. For example, we may be required to lift a heavy load to space for some deflection missions. Will we have appropriate launch capabilities? These studies would also help guide experiments that might be included in upcoming asteroid and comet missions.
7. Examine mission design, political, and policy issues involving the use of nuclear explosives for NEO deflection.
8. Identify and characterize promising non-nuclear means of deflection.
9. Establish the reliability requirement for a NEO deflection mission to insure that the overall probability that the object will impact Earth is less than some defined value (e.g., less than 1 in a million).
10. Develop realistic decision and funding timelines for deflection missions, emphasizing critical political, funding, and mission milestone decisions.
11. Conduct tests of deflection techniques leading to a demonstration of an ability to move an asteroid.

## C. Public Information and Communications

Empirical studies in the social sciences can be brought to bear on public interaction and information aspects of the NEO impact problem, and the resulting information needs to be shared with the physical science and engineering communities. Knowledge of how the public responds to natural disasters (and to warnings of disaster) is obviously useful in developing an overall notification, warning, and response plan.

## Findings:

People often remain calm and perform rationally when confronted with an immediate threat or disaster, and are reasonably tolerant of false alarms as long as they are kept informed. People will seek multiple sources of reliable information should they learn of a warning. If communications about the hazard are to be judged credible by the public, leaders and organizations must begin work now to establish that they are trustworthy. Social science research indicates that a top-down, command-and-control approach to communications is not always best. This finding is relevant to communicating the risk before a real threat is identified, and it is relevant to communicating the risk should a real threat be discovered.

Good science fiction and other popular entertainment have helped and can continue to help raise the awareness of the public to the NEO hazard. However, science fiction based on bad science and engineering might mislead the public and contribute to the "giggle factor" sometimes associated with planetary defense studies. One approach to improving the treatment of the NEO threat and possible mitigation approaches in the media is to make available scientifically accurate presentations, including visuals and explanations, to the media, the press, and leadership.

The potential impact of an asteroid or comet is an international issue, and we need international means to deal with it.

## Recommendations:

12. Apprise the public in an accurate and authoritative way of possible threats and potential actions that might be taken. The Torino Scale and other tools to portray the threat to the public should be reviewed with the goal of improving communications with the public and with decision makers. Warnings must maintain public trust by presenting information using specific, relevant, and consistent terms. It may not be effective to define the threat in probabilistic terms. The public must view the sources of information as objective and trustworthy. The information released should be supported by experts in relevant fields. Social scientists should assist in the development of these protocols.

13. Apply lessons learned from major disasters to help us understand our ability to mitigate a disaster caused by a NEO impact. Deter-

mine whether some NEO impact scenarios are already covered by existing plans for earthquake, tsunami, or other disasters.

14. Bring evidence of previous NEO impacts to the attention of the public to increase awareness that impacts do happen and that the possibility of future impacts should not be ignored.

15. Demonstrate to the public that something can be done about a NEO hazard. A demonstration mission to change the orbit of a non-impacting NEO in a controlled fashion is one way to accomplish this.

## D. Political and Policy Considerations

A deflection mission requires political and policy-level support. A deflection mission with a short time frame would require substantial investment and might require nuclear devices (whose use currently is forbidden in space). Clearly, an authority would need to be in charge of the effort, and preparations for evacuation and other mitigation efforts would need to be made in the case that the deflection effort was unsuccessful. There would be substantial issues related to international participation in a deflection effort and possible disaster mitigation.

## Findings:

At present, there is no formal chain of responsibility for planetary defense. Potential applications of existing and new projects to possible planetary defense use are not currently explored or encouraged. Outside of NASA, there are no established guidelines or protocols for notifying authorities or the public of impact possibilities or even for identifying where such notices should go. No detailed studies of potential disaster scenarios related to NEO impacts have been conducted. Finally, governments have a legal obligation to institute effective plans to prevent this type of harm, and failure to take prompt and meaningful action could result in monetary liability or violation of public international and domestic law.

## Recommendations:

16. Explore political and policy-level decisions and decision timelines for various deflection scenarios (short-term, long-term, nuclear, non-nuclear). Assess potential public and government concerns and

responses to a potential threat and subsequent deflection effort. Consider how a deflection effort would be managed in the face of public expectations and uncertainty.

17. Find (or create) an organizational/governmental home within the U.S. government for the NEO issue. NASA, DoD, DoE, FEMA, DHS, and perhaps a newly formed Space Corps of Engineers based on the Army Corps of Engineers model have been suggested as possibilities. In June 2000, AIAA[4] recommended establishment of an "interagency office charged with dealing with all aspects of Planetary Defense" and further recommended that a "senior level inter-agency working group be formed to define the appropriate makeup and reporting structure of the planetary defense organization, develop a roadmap leading to its implementation, and procure funding for its support." A subsequent AIAA Position Paper[5] reiterated this recommendation.

18. Assess which lessons can be learned from major disasters that would apply to a NEO impact disaster. We need a systematic and thorough review of how the substantial literature on disasters and risk communications can inform NEO-related policy.

19. Begin a dialog among nations and international institutions to characterize the challenges implicit in worldwide planning and execution of future deflection missions.

20. Develop contingency plans and processes for NEO mitigation. Understanding and solving policy issues before they need to be invoked would greatly enhance our ability to mount a successful deflection mission.

## 3. Summary

This conference is the first of a series on the threat posed by Near Earth Objects, possible techniques and missions for deflecting an oncoming object, and political, policy, and disaster-preparedness issues associated with NEO deflection. The conference produced several recommended actions, the foremost being that we need to: 1) begin trust-building efforts so that claims that the

---

[4]Edward Tagliaferri, "Dealing with the Threat of an Impact of an Asteroid or Comet on Earth: The Next Step," AIAA Position Paper, June 2000.
[5]Edward Tagliaferri, Warren Greczyn, and Lawrence Cooper, "Addressing the Comet and Asteroid Impact Threat: A Next Step," AIAA Position Paper, July 2002.

NEO hazard is important will be considered credible by the public, even though we recognize that the probability of a disastrous impact is small; 2) increase our efforts to detect threatening objects and to determine the detailed physical and compositional properties of NEOs; and 3) move forward on means to deflect a threatening object.

A key recommendation, consistent with previous AIAA Position Papers, is that a chain of responsibility be clearly and publicly defined for detecting and warning the public of threats, and mitigating those threats. These threats are real, and efforts to coordinate information and activities related to detecting and mitigating them should begin now.

Considerable work needs to be done to ensure that threats can be detected early and that the means available for deflection are known to be effective.

Future impacts by comets and asteroids are a certainty. Such impacts could have severe consequences—even ending civilization and humanity's existence. Life on Earth has evolved to the point where we can mount a defense against these threats. It is time to take deliberate steps to assure a successful defensive effort, should the need arise.

## Primary Contributors

The individuals below contributed to the development of this document by presenting their ideas at the conference or by participating directly in the document's preparation. The findings and recommendations are based on discussions held during the meeting and are believed to represent consensus opinions of those in attendance; however, some may not reflect the opinions of all individuals listed below or all attendees. Findings and recommendations are personal opinions and are not intended to reflect the opinions of listed organizations.

| | |
|---|---|
| William Ailor* | The Aerospace Corporation (Conference Chairman) |
| Mark Barrera | The Aerospace Corporation |
| Ivan Bekey | Bekey Designs |
| Michael Belton | Belton Space Exploration Initiatives |
| Dennis Byrnes | NASA Jet Propulsion Laboratory |

---

*William Ailor, The Aerospace Corporation, PO Box 92957, Los Angeles, CA 90009-2957, (310) 336-1135. william.h.ailor@aero.org.

| | |
|---|---|
| William E. Burrows | New York University |
| Andrea Carusi | European Space Agency |
| Clyde Chadwick | The Aerospace Corporation |
| Clark Chapman | Southwest Research Institute |
| Steve Chesley | NASA Jet Propulsion Laboratory |
| Paul Chodas | NASA Jet Propulsion Laboratory |
| Lee Clarke | Rutgers University |
| Richard Davies | Western Disaster Center |
| David Dearborn | Lawrence Livermore National Laboratory |
| Fadi Essmaeel | Office of Congressman Dana Rohrabacher |
| George Friedman | University of Southern California |
| Andreas Galvez | European Space Research and Technology |
| Kevin Greene | Civil Engineer, Frontier Consultants |
| Al Harrison | University of California/Davis |
| Kathleen Hollingsworth | Office of Congressman Dana Rohrabacher |
| Keith Holsapple | University of Washington |
| Lindley Johnson | Planetary Science Institute |
| John Logsdon | George Washington University |
| Ed Lu | NASA Johnson Space Center |
| David Lynch | The Aerospace Corporation |
| Jay Melosh | University of Arizona |
| David Morrison | NASA Ames Research Center |
| Oliver Morton | Author |
| Larry Niven | Author |
| Steve Ostro | NASA Jet Propulsion Laboratory |
| Glenn Peterson | The Aerospace Corporation |
| Dan Poston | Los Alamos National Laboratory |
| Jakub Ryzenko | Polish Space Office |
| Dan Scheeres | University of Michigan |
| Rusty Schweickart | B612 Foundation |
| Evan Seamone | Attorney |
| Oleg Shubin | Russian Federal Nuclear Center |
| Vadim Simonenko | Russian Federal Nuclear Center |
| Geoffrey Sommer | RAND |
| Grant Stokes | MIT Lincoln Laboratory |
| Lee Valentine | Space Studies Institute |
| Harvey Wichman | Claremont McKenna College |

| | |
|---|---|
| Simon P. Worden | Congressional Fellow |
| Donald Yeomans | Jet Propulsion Laboratory |

## List of Attendees

| Name | Affiliation |
|---|---|
| Paul Abell | NASA-JSC |
| Robert Adams | NASA-MSFC |
| William Ailor | The Aerospace Corporation |
| Hi Anderson | United Press International |
| Eric Asphaug | |
| Mark Barrera | The Aerospace Corporation |
| Nella Barrera | University of California |
| Ivan Bekey | Bekey Designs |
| Michael Belton | Belton Space Exploration Initiatives, LLC |
| Shmuel Ben-Shmuel | The Aerospace Corporation |
| Jim Benson | Space Development |
| Robert Bernstein | Seaspace Corporation |
| Rick Binzel | MIT |
| John Boylan | Cassidy & Association |
| William Brown | Brown Research |
| James Burke | The Planetary Society |
| William Burrows | New York University |
| Dennis Byrnes | NASA-JPL |
| Eric Campbell | The Aerospace Corporation |
| Jonathan Campbell | NASA-MSFC |
| Andrea Carusi | The SpaceGuard Foundation |
| Claudio Casacci | Alenia Spazio |
| James Cazamias | LLNL |
| Clyde Chadwick | The Aerospace Corporation |
| David Chandler | The B54New Scientist |
| Franklin Chang-Diaz | NASA-JSC |
| Clark Chapman | Southwest Research Institute |
| A.C. Charania | SpaceWorks Engineering, Inc. (SEI) |

| | |
|---|---|
| Steven Chesley | NASA-JPL |
| Vladimir Chobotov | The Aerospace Corporation |
| Paul Chodas | NASA-JPL |
| Jerry Chodil | Ball Aerospace |
| Lee Clarke | Rutgers University |
| Sean Collins | University of California |
| William Cooke | Computer Science Corporation |
| Guy Copper | Sonic Development Laboratories |
| Tracie Crane | Sverdrup/Qualis |
| Leonard David | Space.com |
| Richard Davies | Western Disaster Center |
| David Dearborn | Lawrence Livermore National Laboratory |
| Lawrence Doan | Self-employed |
| Dan Durda | Southwest Research Institute |
| Fadi Essmaeel | REDIS |
| Gary Evans | Science Photo Library |
| Jennifer Evans | MIT Lincoln Laboratory |
| David Farless | NASA-JPL (Retired) |
| Karin Feldman | The Aerospace Corporation |
| Tracy Feltes | Iowa State University |
| George Fenimore | |
| Steve Fluty | Associated Press |
| Francis French | Reuben Fleet Science Center |
| Louis Friedman | The Planetary Society |
| Joseph Frisbee | United Space Alliance |
| John Fujita | The Aerospace Corporation |
| Andres Galvez | European Space Agency—ESTEC |
| Donald Gennery | NASA-JPL (Retired) |
| Thomas Gillon | Department of National Defense |
| Jon Giorgini | NASA-JPL |
| David Glackin | The Aerospace Corporation |
| James Goetz | Lockheed Martin Space System |
| Thomas Goodey | The Flying Kettle Project |
| Kevin Greene | Civil Engineer, Frontier Consultants |
| Peter Gural | SAIC |
| Richard Haase | Space Frontier Foundation |
| Alan Harris | Space Science Institute |
| Albert Harrison | University of California |

| | |
|---|---|
| Mike Hearn | |
| Kathleen Hollingsworth | U.S. Congress |
| Keith Holsapple | University of Washington |
| Alan Holt | NASA-JSC |
| Michael Hout | NASA |
| Piet Hut | Institute for Advanced Study |
| Joseph James | Creative Communications |
| Lindley Johnson | NASA |
| Scott Johnson | Department of National Defense |
| John Junkins | Texas A&M University |
| Ralph Kahle | German Aerospace Center |
| Linda KayBunnell | NASA/LARC |
| Hawkins Kirk | Scientific Applications Research Assoc. |
| Kurt Klaus | The Boeing Company/NASA System |
| David Komm | Boeing Electron Dynamic Device Inc. |
| Bryan Laubscher | Los Alamos National Laboratory |
| John Logsdon | George Washington University |
| Edward Lu | NASA |
| David Lynch | The Aerospace Corporation |
| Claudio Maccone | International Academy of Astronautics |
| Lauren Mahoney-Hopping | University of Florida |
| Francois Martel | Espace Inc. |
| Daniel Mazanek | Spacecraft and Sensors Branch |
| H. Jay Melosh | University of Arizona |
| Ching Meng | The John Hopkins University Applied Physics Laboratory |
| P. K. Menon | Optimal Synthesis INC |
| "Mitch" William Mitchell | NEO Safety International |
| Daniel Molina | NEO Safety International |
| Darryl Moon | MWOA-TIE |
| David Morrison | NASA-Ames Research Center |
| Oliver Morton | The APSE |
| Mason Mulhollad | University of North Dakota |
| Don Nelson | NASA (Retired) |
| Larry Niven | Author |
| Guy Norris | Flight International |
| Scot Osburn | The Aerospace Corporation |
| Steve Ostro | NASA-JPL |

| | |
|---|---|
| Dee Pack | The Aerospace Corporation |
| George Paulikas | The Aerospace Corporation |
| Glenn Peterson | The Aerospace Corporation |
| Frank Pinkney | Department of National Defense |
| Tara Polsgrove | NASA-MSFC |
| David Poston | |
| Rang Ranganathan | Royal Military College |
| Timothy Roberts | SAIC |
| Jakub Ryzenko | Polish Space Office |
| Daniel Scheeres | The University of Michigan |
| Natalia Scherbinskaya | Ministry of Atomic Energy—Russia |
| Phillip Schwartz | The Aerospace Corporation |
| Russell Schweickart | B612 Foundation |
| Evan Seamone | United States Army |
| Vadim A. Simonenko | Ministry of Atomic Energy—Russia |
| Geoffrey Smit | The Aerospace Corporation |
| Dave Smith | The Boeing Company |
| Geoff Sommer | RAND Corporation |
| Laura Speckman | The Aerospace Corporation |
| Grant Stokes | MIT |
| Edward Tagliaferri | The Aerospace Corporation |
| Peter Tennyson | JHU/APL |
| Lee Valentine | Space Studies Institute |
| JeanLuc Verant | ONERA Space System |
| Roger Walker | QinetiQ |
| Henry Wedaa | California Hydrogen Business Council |
| Rob Whiteley | USAF/SMC |
| Harvey Wichman | Claremont McKenna College |
| Bobby Williams | KinetX Inc. |
| Alan Willoughby | Asteroid Al |
| Robert Wood | Boeing (Retired) |
| Donald Yeomans | NASA-JPL |
| Alan Zucksworth | U.S. Air Force/ASC/ENFT Aerospace |

# NOTES

Each note ends with a page number indicating where in the text of *The Survival Imperative* the topic is discussed.

## 1: Hell on Earth

Nuclear Detonation Detection System: Correll, "National Security Implications," 17.

Fire damage: Eden, *Whole World on Fire*, passim, 29.

## 2: Let There Be Light

"SPACEGUARD": Clarke, *Rendezvous with Rama*, 32.

Tyson's definition: Tyson, *Sky Is Not the Limit*, 169, 33.

150-meter-plus impacts: *Spaceguard Survey*, 34.

Threshold size for global catastrophe: ibid., 35.

Science Definition Team: *Study to Determine the Feasibility*, 35.

Optical versus infrared: ibid., 36.

Soot and warming: "Black Soot and Snow: A Warmer Combination," NASA Home Page, December 23, 2003; and "Soot Is Cited as Big Factor in Global Warming," *New York Times*, December 25, 2003, 37.

Inactive addresses: Dellavalle et al., "Going, Going, Gone," 787, 38.

Lindsey on the broken Earth: Lindsey, *Late Great Planet Earth*, 42.

Fright or deliverance: ibid., 42.

Parody in *Nature*: Hartmann, "Paradigm and the Pendulum," 43.

Religious fanatics: Drosnin, *Bible Code II*, 218, 44.

bin Laden: ibid., 230, 44.

Arafat's assassination: ibid., 100, 44.

Rips's holocaust: ibid., 236, 44.

Debunking Drosnin: Shermer, "Codified Claptrap," 35, 45.

*Moby-Dick* and assassinations: www.nmsr.org/biblecod.htm, 45.

Rees on protecting intelligent life: Rees, *Our Final Hour,* 45.

## 3: Target Earth

Vajpayee and war: "The Kashmir Brink," *New York Times,* June 20, 2002, 49.

Casualties and consequences: "12 Million Could Die at Once in an India-Pakistan Nuclear War," *New York Times,* May 27, 2002, 49.

Worden's warning: "Incoming," *New York Times,* October 13, 2002, 49.

Carolyn's finding the comet: Petersen and Brandt, *Hubble Vision,* 116, 51.

Representative Brown's prediction: "When Worlds Collide: A Threat to the Earth Is a Joke No Longer," *New York Times,* August 1, 1994, 51.

Harris and Morrison on cost-benefit: *NEO News* (online from the Ames Research Center), October 26, 2001, 52.

The K-T event and multiple hits: "New Theory on Dinosaurs: Multiple Meteorites Did Them In," *New York Times* (Science Times), November 5, 2002, 53.

"Threshold Size for Global Catastrophe": *Spaceguard Survey,* 10, 53.

Ranking hazards: ibid., 49–50, 54.

Frequency of impacts: "The Cosmic Time Bomb Waiting to Go Off: Government Urged to Establish Safeguard Against Deadly Comets," *Guardian,* September 16, 2000, 54.

"end civilization as we know it": *Spaceguard Survey,* 7, 54.

Survey Program details: ibid., 50–52, 54.

EUNEASP: Hahn et al., "The EUNEASCO Project: A European NEO Search, Follow-up, and Physical Observation Programme," in Remo, *Near-Earth Objects,* 27, 55.

The iridium theory: Yeomans, *Comets,* 354, 55.

*Endeavour* and Yucatán: Perkins, "Killer Crater," *Science News,* 163, 55.

Purpose of the UN meeting: Remo, *Near-Earth Objects,* ix, 56.

NEAT: Helin, Pravdo, Rabinowitz, and Lawrence, "Near-Earth Asteroid Tracking," in Remo, *Near-Earth Objects,* 10, 56.

Morrison on YB5: *NEO News,* January 9, 2002, 58.

YB5 media coverage: ibid., 60.

2002 EM7: "Asteroid Buzzes Earth from 'Blind Spot,'" NewScientist.com, March 15, 2002, 60.

2002 MN: "Too Close for Comfort: Asteroid Passed within 75,000 Miles of Earth," Space.com, June 20, 2002, 60.

Morrison's lament: *NEO News*, June 24, 2002, 60.

1950 DA: Giorgini et al., "Asteroid 1950 DA's Encounter," 132–36, 61.

Yarkovsky effect: Spitale, "Asteroid Hazard Mitigation," 77, 62.

Dusting it with soot: "Encounter with an Asteroid," *New York Times*, April 8, 2002, 62.

Giorgini on 1950 DA: Braun, "Is a Large Asteroid Headed," 62.

1950 DA hits the Atlantic: Ward and Asphaug, "Asteroid Impact tsunami," F6, 62.

Crying wolf: Chapman, Durda, and Gold, "The Comet/Asteroid Impact Hazard," 7, 62.

Midwest explosion: "Meteorite Shower in Illinois," *NEO News*, March 28, 2003. The Auckland hit: "Meteorite Smashes Through Home," AOL News (Reuters), June 13, 2004, 64.

Italian earthquake: "A Town in Italy Loses Its Future to Tiny Coffins," *New York Times*, November 2, 2002, 64.

"A miss": "Quake Forecasting Booms, but Results Lag," *New York Times*, May 13, 2003, 65.

Moving the Iranian capital: "With Capital at High Risk of Quakes, Iran Weighs Moving It to a Safer Place," *New York Times*, January 6, 2004, 65.

Clarke communication: To B. J. Peiser, December 29, 2004, 66.

Krakatoa: Richard Ellis, "The Island That Went Straight Up," *New York Times Book Review*, April 20, 2003, 9, 67.

The Tambora eruption: Evans, "Blast from the Past," 54–57, 68.

"It is no exaggeration": Rhodes, *Making of the Atomic Bomb*, 728, 68.

Unspeakable suffering: ibid., 742, 68.

"Unacceptable damage": Dyson, *Weapons and Hope*, 239, 69.

Schlesinger and McNamara: Rees, *Our Final Hour*, 26–27, 69.

Bundy on the crisis: Bundy, *Danger and Survival*, 453, 70.

Number of nukes and budget: Cochran, Arkin, and Hoenig, *U.S. Nuclear Forces*, 2, 70.

Overkill and Hiroshima: Pringle and Arkin, *S.I.O.P.*, 103, 71.

Eisenhower on annihilation: "Discussion at the 257th Meeting," 9, 11, 71.

Risk a one in six chance: Rees, *Our Final Hour*, 28, 71.

New York City and burn victims: Keeney, *Doomsday Scenario*, 81, 72.

Economic and governmental chaos: ibid., 66, 72.

Protecting art: ibid., 119, 72.

Survival of the rabbits and the poisoning: Shute, *On the Beach*, 260, 320, 73.

Schell on nuclear destruction: Schell, *Fate of the Earth*, p. 181, 74.

Poisonous snakes: "C.I.A. Nominee Wary of Budget Cuts," *New York Times*, February 3, 1993, 75.

Abkhazia: Stone, "Fears Grow of Nuclear Brain Drain," 1498, 76.

Keller's observations: Keller, "Thinkable," 50–51, 77.

About six hundred tons: Stone, "Nuclear Trafficking," 1632–36, 77.

Cesium-137: "Police in Thailand Seize Radioactive Material," *New York Times*, June 14, 2003, 78.

Bearden on terrorism: "You Cut the Head, but the Body Still Moves," *New York Times*, March 21, 2004, 78.

Proliferation rundown and Ekeus and Schmid: Stone, "Nuclear Trafficking," 1632–36, 78.

Alibek's story: Henderson, "Public Health Preparedness," in Teich, Nelson, and Lita, *Science and Technology*, 33, 78.

Shchuch'ye: "Russia's Poison Gases," *New York Times*, October 30, 2002, op-ed, 79.

Rees and terror nukes: Rees, *Our Final Hour*, 42, 79.

Varmus on smallpox: "Borne on the Winds of War," *New York Times Book Review*, November 10, 2002, 79.

Smallpox vaccinations: *New York Times*: "Proposal to Test Smallpox Vaccine in Young Children Sets Off Ethics Debate," November 5, 2002; "Pentagon Plans Smallpox Shots for up to 500,000," October 12, 2002; and "White House Debate on Smallpox Slows Plan for Wide Vaccination," October 13, 2002, 80.

Clarke on Al Qaeda and Iraq: "Former Terrorism Official Faults White House on 9/11," *New York Times*, March 22, 2004, 81.

The Buddhas and the museum in Kabul: Lawler: "Afghanistan's Challenge," 1195; "Buddhas May Stretch Out," 1204; "Then They Buried," 1202, 82.

Putin's warning: "Putin Vows Hunt for Terror Cells Around the World," *New York Times*, October 29, 2002, 83.

Nuclear Posture Review: "U.S. Works Up Plan for Using Nuclear Arms," *Los Angeles Times*, March 9, 2002; and William M. Arkin, "Secret Plan Outlines the Unthinkable," *Los Angeles Times*, March 10, 2002, 84.

*Times* editorial: "America as Nuclear Rogue," March 12, 2002, 84.

Defeating the NPR: "Panel Rejects Nuclear Arms of Small Yield," *New York Times*, May 15, 2003, 84.

Diabolical: "Cold War Long Over, Bush Administration Examines Steps to a Revamped Arsenal," *New York Times*, May 29, 2003, 84.

Kurzweil: *Age of the Spiritual Machine*, 141–42, 86.

Drexler on nanotechnology: Phoenix and Drexler, "Safe Exponential Manufacturing," passim, 86.

Joy on thinking machines: Joy, "Why the Future Doesn't Need Us," 239, 86.

Rees on Wells: Rees, *Our Final Hour*, 9–10, 87.

Rees's warning: ibid., 10–11, 87.

"Just a fanatic": ibid., 3, 87.

Hawking and space: "Scientist: Humans Doomed Without Space Colonies," Reuters, October 16, 2001.

Hawking and greenhouse: "Stephen Hawking Fears the Human Race May Not Survive Another Millennium," Associated Press, September 30, 2000, 88.

Truman and the bomb: Truman, *Years of Trial and Hope*, 395, 89.

Radioactive cone: Baker, *Shape of Wars to Come*, 160, 89.

Nostradamus: de Fontbrune, *Nostradamus*, passim, 89.

Doomsday: Wilson, *Mammoth Book of Nostradamus*, passim, 90.

The "arrow": Welch, *Comet of Nostradamus*, 13, 90.

Rees on space: Rees, *Our Final Hour*, 170, 90.

Hawking the optimist: "Scientist: Humans Doomed," 90.

## 4: The Once and Future Space Program

*True History*: Cain, *Luna Myth*, 154; and Crouch, "To Fly to the World," 10, 92.

"For myself ": Hale, "Brick Moon" (part one), 455, 93.

Tsiolkovsky's social problems: Riabchikov, *Russians in Space*, 92, 93.

Message from space on April 10, 2017: Tsiolkovsky, *Beyond the Planet Earth*, 107, 94.

Von Braun's report: McGovern, *Crossbow and Overcast*, 147–48, 96.

Pendray and Martial's prediction: Ratcliff, *Science Year Book of 1945*, 156–57, 96.

Kaempffert's prediction: Kaempffert, *Science Today and Tomorrow*, 143, 97.

Clarke's pressure domes: Clarke, *Exploration of Space*, 145, 98.

Ryan on the stages to space: Ryan, *Across the Space Frontier*, xi–xii, 99.

Ryan on the Russians: ibid., xiii–xiv, 99.

Wooden rubles: Interview with the author, 100.

Tikhonravov's report: Tikhonravov, "Creation of the First Artificial Earth Satellite," 207–8, 100.

Provisions of the Space Act: Committee on Commerce, *National Aeronautics and Space Act of 1958*, 1, 102.

Allen's testimony: Logsdon, *Decision to Go*, 25, 104.

Sorensen on JFK and the Moon: Sorensen, *Kennedy*, 525, 105.

JFK's Moon speech: Logsdon, *Decision to Go*, 128, 105.

Logsdon on Kennedy: ibid., 27–28, 106.

Kennedy's change of heart: Logsdon, *Exploring the Unknown*, 381, 107.

Sagan on the Moon: Sagan, *Pale Blue Dot*, 206, 107.

Webb's predicament: McDougall, *Heavens and the Earth*, 420–21, 108.

LBJ and Webb: Levine, *Managing NASA*, 257n, 109.

Von Braun's prediction: Von Braun, "Exploration to the Farthest Planets," 1:39–42, 109.

Poll results: Roger D. Launius, "Why Go to the Moon? The Many Faces of Lunar Policy" (paper presented at the American Astronautical Society meeting, Greenbelt, Md., March 17, 2004), 110.

Agnew and sustaining interest: *Post Apollo Space Program*, 5, 110.

The PSAC's five programs: *The Space Program in the Post-Apollo Period*, 14, 110.

Moon shots in stride: Ezell, *NASA Historical Data Book*, 3:54, 111.

"moondoggle": Lewis, *Voyages of Apollo*, 150, 111.

No ticker-tape parades: Pathway to the Planets meeting sponsored by NASA's Office of Exploration, May 31–June 1, 1989, 113.

NASA's budget decline: Ezell, *NASA Historical Data Book*, 3:61, 113.

Finger's observation: Levine, *Managing NASA*, 98–99, 114.

Space Task Group recommendation: Ezell, *NASA Historical Data Book*, 3:113–14, 116.

GAO report: "Analysis of Cost Estimates for the Space Shuttle and Two Alternate Programs," 1, 116.

Clauser, Hunter, and Mueller: Heppenheimer, *Space Shuttle Decision*, 245–46, 117.

Weeks shaking his head: "In Harsh Light of Reality, the Shuttle Is Being Re-evaluated," *New York Times*, May 14, 1985, 119.

Reagan and the station: Mark, *Space Station*, 195, 120.

Keyworth's plan: ibid., 185, 120.

Dissenting views on the station: ibid., 136–39, 145, 167, 121.

*Times* editorial: "The Wrong Stuff," *New York Times*, November 21, 1984, 121.

Commentary piece: "The Space Station Is Losing Friends," *Science*, October 19, 1990, 364, 122.

Bush's scaled-down station: Reichhardt, "Cutbacks 'Will Cripple,'" *Nature*, 263; and "Cuts Lessen Space Station's Value to Science, Report Says," *New York Times*, September 20, 2002, 123.

Knowing they were doomed: "Crew of Columbia Survived a Minute After Last Signal," *New York Times*, July 16, 2003, 123.

Settles and Ronney: "In Wake of Columbia Disaster, Scientists Question the Value of Shuttle Flights," *New York Times*, February 24, 2003, 124.

McDonald's testimony: "NASA Records in Disarray, Study Leader Tells Board," *New York Times*, March 7, 2003, 124.

Fatigue factor: "Report Criticizes NASA's Shuttle Maintenance and Predicts Further Fatal Accident," *New York Times*, July 15, 2003, 125.

Safety Center: "NASA, in Response to *Columbia* Panel, Plans an Agencywide Safety Center," *New York Times*, July 16, 2003, 125.

Complacency: "Shuttle Investigator Faults NASA for Complacency over Safety," *New York Times*, July 17, 2003, 125.

O'Keefe's testimony: "NASA Official Says Agency Faces a Drain on Its Talent Pool," *New York Times*, March 7, 2003, 125.

Erikson and the *ISS*: Erikson, "Why Space Stations Are Crucial," 23, 126.

Letter to Congress: "An Open Letter to Congress on Near Earth Objects," appendix A, 128.

Space Foundation board: "Space Foundation Urges Space Faring Nations to Address Near Earth Object Impact Threat," Colorado Springs, *Space Foundation*, November 24, 2003, 128.

Betts and Blair: "Nuclear War Strategists Rethink the Unthinkable," *New York Times*, January 19, 2003, Week in Review, 130.

U-2s and Dimona: Burrows and Windrem, *Critical Mass*, 300, 130.

Lunar base study: Eckart, *Lunar Base Handbook*, passim, 130.

## 5: A Beehive Called Earth

The Rand study: *Preliminary Design of an Experimental World-Circling Spaceship*, 1–2, 135.

Tikhonravov and Korolyev: McDonald, *Corona*, 86, 135.

Tikhonravov on *Sputnik*: Tikhonravov, "Creation of the First Artificial Earth Satellite," 207–8, 136.

Korolyev's boys: ibid., 208, 136.

The Soviet handbook: Shternfeld, *Soviet Space Science*, passim, 136.

First successful recovery: Ruffner, *Corona*, 22, 137.

Gagarin on television: Wheelon, "Technology and Intelligence," 251, 138.

Border and embassy sensors: "Grabbing the Nettle," *New York Times*, August 1, 2003, op-ed; and "Russia Helped U.S. on Nuclear Spying Inside North Korea," *New York Times*, January 20, 2003, 139.

Satellites against Al Qaeda: *9/11 Commission Report*, 127; and "Kidnapping of bin Laden Was Rehearsed in '98 but Scrapped, 9/11 Report Says," *New York Times*, July 26, 2004, 139.

Tracking Khan's weapons: "A Tale of Nuclear Proliferation: How Pakistani Built His Network," *New York Times*, February 12, 2004, 139.

Satellites over Yongbyon: "North Korea Hides New Nuclear Site, Evidence Suggests," *New York Times*, July 20, 2003, 139.

Commercial satellite images: "North Korea Begins to Reopen Plant for Processing Plutonium," *New York Times*, December 24, 2002, 140.

Yongbyon close-ups: "Satellites Said to See Activity at North Korean Nuclear Site," *New York Times*, January 31, 2003, 140.

Youngdoktong: "C.I.A. Said to Find Nuclear Advances by North Koreans," *New York Times*, July 1, 2003, 140.

"Spy satellites show": "U.S. Sees Quick Start of North Korea Nuclear Site," *New York Times*, March 1, 2003, 140.

Smuggled nuclear equipment: "U.N. Warns of Possible Nuclear Thefts in Iraq," *New York Times*, April 16, 2004, 140.

Boeing's problem: "Boeing Lags in Building Spy Satellites," *New York Times*, December 4, 2003, 140.

Commercial satellite dangers: Miller, Stocker, and Martel, "Commercial Space Systems," 11, 141.

Future surveillance systems: Teets, "Challenges for National Reconnaissance," 6, 141.

CAC: Baclawski and Nath, "Civil Applications Committee's Role," 13–15, 142.

Convoys from Iraq to Syria: "Iraqis Removed Arms Material, U.S. Aide Says," *New York Times*, October 29, 2003, 142.

Forced labor camps: "Rights Group Exposes Conditions in North Korean Prison Camps," *New York Times*, October 22, 2003, 143.

Hidden facilities: "North Korea Hides New Nuclear Site," 143.

North Korean admission and threat: "North Korea Says It Now Possesses Nuclear Arsenal," *New York Times*, April 25, 2003; "North Korea Says It Has Made Fuel for Atom Bombs," *New York Times*, July 15, 2003; and "North

Korea Says a U.S. Attack Could Lead to a Nuclear War," *New York Times*, March 3, 2003, 143.

Scuds to Yemen: "Scud Missiles Found on Ship of North Korea," *New York Times*, December 11, 2002; and "Reluctant U.S. Gives Assent for Missiles to Go to Yemen," *New York Times*, December 12, 2002, 144.

Routine monitoring and drugs: "North Korea, Protesting New Ship Inspections, Suspends Its Passenger Ferry Link with Japan," *New York Times*, June 9, 2003, 145.

UN bombing: "Another Bombing at the U.N. in Baghdad," *New York Times*, September 23, 2003, 145.

Car bomb imagery: "Car Bomb in Central Baghdad Kills 27 near Anniversary of War's Start," *New York Times*, March 18, 2004, 145.

Private space imaging: "U.S. Allows Bird's-Eye View," *Wall Street Journal*, March 21, 2003, 146.

Mrs. Crider's problem: "Line Dispute Would Make Connecticut's Gain Rhode Islander's Pain," *New York Times*, April 3, 2004, 146.

Tsiolkovsky and Earth observation: Winter, "Camera Rockets," 73, 147.

Pentagon fighting for EROS: Webb, "Strange Career of Landsat," 17, 147.

Early history of ERTS: Ezell, *NASA Historical Data Book*, 3:335–36, 147.

ERTS resolution: Short et al., *Mission to Earth*, 439, 148.

Corona resolution: McDonald, *Corona*, 306, 148.

Fiske Creek fire: Short et al., *Mission to Earth*, 7, 149.

Houston imagery: ibid., 11, 149.

New England land use: ibid., 9, 149.

Genetically modified corn: Knight, "US Reflects on Flying Eye," 112, 150.

Letter to *Nature*: "Lack of Concern Deepens the Oceans' Problems," *Nature*, August 14, 2003, 151.

Overfishing: Pauley and Watson, "Last Fish," 43–47; and "Panel Calls for Sea Change," 577, 151.

Sachs: "New Technological Approaches—Not Just Better Policies Required," *Update*, 6, 151.

Landsat's uses: Short et al., *Mission to Earth*, 4, 151.

Spotting ocean debris: "Satellites and Airborne Searches Spot Harmful Ocean Debris," *New York Times*, October 14, 2003, 152.

Deforestation in the Amazon: "Rain Forest Is Losing Ground Faster in Amazon, Photos Show," *New York Times*, June 28, 2003, 152.

Shrinking Arctic ice: "Whither Arctic Ice?" *Science*, August 2002, 1491, 152.

Ward Hunt Ice Shelf: "Huge Ice Shelf Is Reported to Break Up in Canada," *New York Times*, September 23, 2003, 152.

Browne on ice: "Under Antarctica, Clues to an Icecap's Fate," *New York Times*, October 26, 1999, Science Times, 153.

Disintegration of the shelf: "Large Ice Shelf in Antarctica Disintegrates at Great Speed," *New York Times*, March 20, 2002, 153.

The hurricane of 1938: "Remembering the Great Hurricane of '38," *New York Times*, September 21, 2003, 154.

Cooperative forest-fire effort: "From Way on High, Help in Fighting Forest Fires," *New York Times*, August 19, 2003, 156.

Aura: "NASA's New Eye on Sky to Watch Earth's Ozone," *New York Times*, July 6, 2004, 156.

Natural Hazards and Earth Observatory: E-mail from David Herring to the author, August 8, 2003, 157.

Pre- and post-blackout satellite imagery: "Across the Region, Waiting for a Flicker: A Long Day's Journey into Light," *New York Times*, August 16, 2003, 157.

Global data sharing: "Summit Pledges Global Data Sharing," *Science*, April 30, 2004, 158.

Lowman on tectonics: Lowman, "Space Exploration and Tectonics," 9, 159.

Equator-weight problem: Perkins, "Mapping with Grace," *Science News*, January 4, 2003, 159.

GRACE: "Grace on Track to Map Earth's Shifting Mass," *SpaceNews*, September 23, 2002; and Tapley, "Early Results from the Gravity Recovery and Climate Experiment" (presentation at the AAAS annual meeting, 2003), 160.

Quakesat: "Satellites Aim to Shake Up Quake Predictions," *Nature*, July 31, 2002, 478, 160.

STG and Jimmy Carter for applications: Ezell, *NASA Historical Data Book*, 3:236, 161.

Clarke on comsats: Edelson, "Communication Satellites," 95, 161.

Helping Afghanistan: "Space Scientists Support Satellites as Aid for Afghanistan," *Nature*, 620, 162.

JFK on comsats: Edelson, "Communications Satellites," 96–97, 162.

Telstar: ibid., 101, 162.

Nukes in orbit: Dupont, "Nuclear Explosions in Orbit," 102, 165.

GOES launches: "NOAA's GOES-9 Is Lined Up for Storm-Tracking Duty," *SpaceNews*, January 29–February 4, 1996, 166.

Nelson: "Earth Observation Officials Grapple With Growing Demand," *SpaceNews*, October 14, 2002, 167.

Communication and meteorological data: Yenne, *Encyclopedia of US Spacecraft*, 168.

GPS III: Ashley, "Next Generation GPS," 34, 169.

Solar panel problems: Deckard, "Technology for a Better Future," 130–31, 171.

Solar power and the parasol: Hoffert et al., "Advanced Technology Paths," 984, 171.

Twenty-four thousand objects: Verger, Sourbes-Verger, and Ghirardi, *Cambridge Encyclopedia of Space*, 55, 171.

Erikson on debris: Erikson, "Growing Space" (manuscript), 8–2, 171.

Shrapnel: Erikson, "Spaceways System," 172.

Turner on collisions: E-mail of September 8, 2003, 172.

Fudge: Fudge, "LEO Constellation Orbital Debris," 9, 172.

Johnson and debris: " 'No Littering' Plea Extended to Space Junkyard," Space.com, May 17, 2002, 172.

Radioactive droplets: "Havoc in the Heavens: Soviet-Era Satellite's Leaky Reactor's Lethal Legacy," Space.com, March 29, 2004, 173.

Skybusses: Erikson, "Growing Space," 8–3, 173.

## 6: The Ultimate Frequent-Flier Program

Cernan's good-bye: Cernan, *Last Man on the Moon*, 337, 174.

Forman's opinion: Forman, "Launch Systems," 175.

"Butterfly on a Bullet": "Decoding *Columbia*: A Detective Story," *Los Angeles Times*, December 21, 2003, 176.

Covey: "Shuttle Flights May Resume Before NASA Culture Changes," *New York Times*, August 8, 2003, 176.

Iceberg of problems: "NASA's Failings Go Far Beyond Foam Hitting Shuttle, Panel Says," *New York Times*, June 7, 2003, 176.

ASAP: "All 9 Members of a NASA Safety Panel Resign," *New York Times*, September 23, 2003, 177.

Student experiments: "Hopes for Experiments Disappear as Disaster Unfolds," *New York Times*, February 2, 2003, 178.

Feynman's warning: *Report of the Presidential Commission*, 2:F-5, 179.

Osepok's warning: Aldrin and Barnes, *Encounter with Tiber,* 189, 179.

Erikson on Nature: Erikson, "Why Space Stations Are Crucial," 23, 179.

Erikson on survival: Erikson, "The Spaceways System," 2, 179.

Kostelnik's remarks: "Future Shuttles May Carry Fewer Astronauts or None at All, NASA Official Says," *New York Times,* March 25, 2003, 180.

Erikson on power: E-mail to the author, August 3, 2003, 181.

The boneyard: Handberg, "Back to the Future," 13–14, 182.

X-43A: "NASA Jet Sets Record for Speed," *New York Times,* November 17, 2004, 183.

Bush's initiative and Hollings's answer: "Bush Backs Goal of Flight to Moon to Establish Base," *New York Times,* January 15, 2004, 184.

Erikson on OSP capabilities: Erikson, "Growing Space" (manuscript), 8–4, 184.

Aldridge panel: "Moon-to-Mars Commission Recommends Major Changes at NASA," Space.com, June 16, 2004; and "NASA Is Urged to Widen Role for Businesses," *New York Times,* June 15, 2004, 185.

Reynolds's view: Reynolds, "Space Frontier Paradigm," 20–21, 186.

Mars mission challenges: Stevens, "Bumpy Road to Mars," 39–40, 187.

Recipe for murder: The author was present.

Melvill's flight: "Manned Private Craft Reaches Space in a Milestone for Flight," *New York Times,* June 22, 2004; "At One Point, 'I Was Deathly Afraid,' New Space Visitor Admits," *New York Times,* June 23, 2004; and "First Private, Manned Craft Achieves Space Flight," AOL News, June 21, 2004, 188.

Rutan's recollections: *60 Minutes* (CBS), November 7, 2004, 188.

Melvill's record and spin: "Private Rocket Recovers, Enters Space," AOL News, September 29, 2004; and "Going Private: The Promise and Danger of Space Travel," Space.com, September 30, 2004, 189.

Ansari X Prize: "The Future of Space Travel," *Wall Street Journal,* June 30, 2004, 189.

Bigelow: Tariq Malik, "Going Private: The Promise and Danger of Space Travel," Space.com, September 30, 2004, 189.

Astronaut wings: "Now Earning Wings, a New Kind of Astronaut," *New York Times,* October 12, 2004, 189.

Branson's plan: "Virgin Territory," 19; *60 Minutes,* November 7, 2004, 189.

"Screwed": *60 Minutes,* November 7, 2004, 189.

People killed in space: ibid, 190.

Aquila: Starcraft Boosters presentations of June 6 and July 22, 2003, and author interview, 190.

Three competing contractors: "NASA Goes Shopping for a Shuttle Successor, Off the Rack," *New York Times,* July 1, 2003, 193.

Kranz on failure: Kranz, *Failure Is Not an Option,* 12, 193.

Aquarius: Turner, "Aquarius Launch Vehicle," 4–5, 194.

Erikson on the *ISS:* Erikson, "Why Space Stations Are Crucial," 23, 195.

Space elevator: "Not Science Fiction: An Elevator to Space," *New York Times,* September 23, 2003, 196.

Erikson on the elevator: E-mail to the author, September 24, 2003, 197.

*Limits to Growth:* Launius and McCurdy, *Imagining Space,* 136–37, 197.

O'Neill on environmental overload: O'Neill, *High Frontier,* 15–17, 198.

O'Neill on Heilbroner: ibid., 20–21, 199.

Mark on O'Neill: Mark, *Space Station,* 52, 199.

O'Neill on Tsiolkovsky: O'Neill, *High Frontier,* 35, 199.

"More surprised": ibid., 56–57, 200.

Risks and dangers: ibid., chapter 7, passim, 202.

Solar flares: "The Biggest Explosions in the Solar System," Science at NASA (Web site), February 6, 2002; "Solar Flares on Steroids," Science at NASA, September 12, 2003; and "The Third Interplanetary Network," IPN3 home page, March 27, 2001, 203.

Lagrange: O'Neill, *High Frontier,* 61–63, 203.

Mass driver: ibid., 70–73; and A. C. Clarke, *Journal of the British Interplanetary Society* 9 (1950), 203.

Island Two: O'Neill, *High Frontier,* 93–94, 203.

Island Three and larger: ibid., 37, 204.

Island Three details: ibid., chapter 5, passim, 204.

The argument against O'Neill: Launius and McCurdy, *Imagining Space,* 142, 205.

Dyson introduction: O'Neill, *High Frontier,* 5–7, 206.

Homesteading asteroids: ibid., 104–14, 206.

## 7: A Treasure Chest on the Moon

Rand and MIT lunar work: Allen, "Early Lunar Base Concepts," 1–2, 209.

Clarke on Moon missiles: Clarke, *Exploration of Space,* 188, 210.

Horizon and Boushey: Allen, "Early Lunar Base Concepts," 1, 4–5, 210.

The Directorate's report: "Military Lunar Base Program," 1:1, 1:6, 210.

Clarke deploring war rockets: Clarke, *Exploration of Space,* 188, 211.

Russians to Mars: "Scientists Chart a Return to the Moon for New Exploits," *New York Times*, December 4, 1984, 211.

Taylor's announcement: Covault, "Manned U.S. Lunar Station," 73, 211.

Van Allen: "Scientists Chart a Return," 212.

Burke: ibid, 212.

*Science* magazine on the symposium: "Asking for the Moon," 948, 212.

*Collier's* on commercial exploitation: Ryan, *Across the Space Frontier*, 122–24, 212.

Liquid oxygen on the Moon: Bulban, "Economic Benefits of Lunar Base Cited," 132.

Prospectors: National Commission on Space, *Pioneering the Space Frontier*, 63, 213.

Ride on mining: Ride, "Leadership and America's Future in Space," 31, 213.

Synthesis Group and mining: Synthesis Group, "America at the Threshold," 26, 213.

Quayle on Mars: CNN interview aired on August 18, quoted in the *San Jose Mercury News*, September 1, 1989, 213.

Lewis on lunar resource development: Lewis and Lewis, *Space Resources*, 194–211, 214.

X-33 and X-34 development: O'Dale, "Establishing an Infrastructure," 45, 215.

"Set up": Author conversation with O'Keefe on December 10, 2003, 216.

Bush and the SEI: Chaikin, "Shoot for the Moon," 45, 216.

Japan's lunar program: "New Technologies Key to Building Permanent Manned Moon Station," *Space News*, October 21–27, 1996, 216.

Indian and Chinese programs: "New Race to the Moon," 724–27, 217.

"Sphere of warfare": "Heading for the Stars, and Wondering if China Might Reach Them First," *New York Times*, January 22, 2004, 217.

Bush's initiative: "Bush Backs Goal of Flight to Moon to Establish Base," *New York Times*, January 15, 2004, 218.

Budget deficits: "Budget Office Forecasts Record Deficit in '04 and Sketches a Pessimistic Future," *New York Times*, January 27, 2004, 218.

AIAA release: "AIAA Responds to President Bush—Puts a Stake in the Ground and Aims for the Stars," AIAA Communications, January 14, 2004, 219.

O'Keefe to Mikulski: "NASA Chief Affirms Stand on Canceling Hubble Mission," *New York Times*, January 29, 2004, 219.

Killing Hubble: "NASA Cancels Trip to Supply Hubble, Sealing Early Doom," *New York Times*, January 17, 2004, 219.

McCurdy and Pike: "Tempered Glee Among Space Experts as the History Books Offer Reasons for Caution," *New York Times*, January 15, 2004, 220.

Ride on the lunar base: Ride, "Leadership and America's Future in Space," 29, 31, 220.

Commission on planetary defense: "Final Report," 3–6, 221.

Chaikin and the Moon: Chaikin, "Shoot for the Moon," 45, 222.

Lunar and Martian lifeboats: Lowman, "Return to the Moon," 1, 222.

Position is everything: Ward and Brownlee, *Rare Earth*, passim, 36.

SETI and dispersal: Lowman, "Return to the Moon," 10–11, 223.

No minerals but helium-3: ibid., 6, 223.

Clarke's lunar base: Clarke, *Promise of Space*, 204, 207, 224.

Saturn V versus shuttle: E-mail from Turner, September 8, 2003, 225.

Goddard's lunar base: Eckart, *Lunar Base Handbook*, 217, 227.

Shevchenko on site selection: Shevchenko, "Lunar Base Siting," 203–4, 227.

Lunar base elements: Eckart, *Lunar Base Handbook*, 247–48, 228.

BLSS: ibid., 405, 228.

Sagan on the library: Sagan, *Cosmos*, 18–19, 229.

Sagan's lament: ibid., 334, 229.

Nearly 10,500 objects: Lawler, "Mayhem in Mesopotamia," 584, 230.

City of lights: "Saving Iraq's Archaeological Past from Thieves Remains an Uphill Battle," *New York Times*, April 4, 2004, 230.

Digitized Library of Alexandria: "Online Library Wants It All, *Every* Book," *New York Times*, March 1, 2003; and "A Temple of Knowledge," *Nature*, October 10, 2002, 231.

Digitized recordings and fossils: "Library of Congress Begins Effort to Protect Recordings," *New York Times*, January 28, 2003; and "In Virtual Museums, an Archive of the World," *New York Times*, January 12, 2003, 231.

Shapiro's reflection: Shapiro, "Backing Up Our Civilization," manuscript for a magazine article, Spring 2003, 233.

ARC: Burrows and Shapiro, "Alliance to Rescue Civilization," 18–21, 234.

Business lunar cache: Turner, "Ultimate in Data Storage," 10–11, 235.

## 8: The Guardians

Wilford on the space program: "Our Future in Space Is Already History," *New York Times*, February 9, 2003, 237.

Bean on space: "Giant Leap to Moon, Then Space Lost Allure," *New York Times*, February 9, 2003, 237.

Wilson's theory: Acceptance speech, Kistler Prize, quoted in *The Next Thousand Years TV Series Project*, 96, 238.

Logsdon's observations: "Amid Inertia and Indecision, Helm Didn't Respond," *New York Times*, August 27, 2003, 239.

Wolfe's letter: "We Mustn't Give Up Space Flight Dreams," *New York Times*, February 9, 1993, 241.

CAPS: Mazanek et al., "Comet/Asteroid Protection System," passim, 243.

B612 Foundation's plan: Schweickart et al., "The Asteroid Tugboat," passim, 244.

Joint projects: Bainbridge, *Goals in Space*, 122, 248.

Harrison on settlements: Harrison, *Spacefaring*, 222, 249.

Schell on life: Schell, *Fate of the Earth*, 154, 249.

Rees on the life imperative: Rees, *Our Final Hour*, 170, 249.

Murray on humanity: Murray, *Navigating the Future*, 10, 11, 250.

# SOURCES

## Books

Aldrin, Buzz, and John Barnes. *Encounter with Tiber.* New York: Warner Books, 1996.

Bainbridge, William Sims. *Goals in Space: American Values and the Future of Technology.* Albany: State University of New York Press, 1991.

Baker, David. *The Shape of Wars to Come.* Cambridge: Patrick Stephens, 1981.

Berkner, Lloyd V., and Hugh Odishaw, eds. *Science in Space.* New York: McGraw-Hill Book Company, 1961.

Bleeker, Johan A. M., Johannes Geiss, and Martin C. E. Huber, eds. *The Century of Space Science.* Kluwer, 2002.

Buchheim, Robert W. *Space Handbook* (Rand Corporation). New York: Random House, 1958.

Bundy, McGeorge. *Danger and Survival: Choices About the Bomb in the First Fifty Years.* New York: Random House, 1988.

Burdick, Eugene, and Harvey Wheeler. *Fail-Safe.* Hopewell, N.J.: Ecco Press, 1962.

Burrows, William E., and Robert Windrem. *Critical Mass: The Dangerous Race for Superweapons in a Fragmenting World.* New York: Simon & Schuster, 1994.

Cain, Kathleen. *Luna Myth & Mystery.* Boulder: Johnson Books, 1991.

Calder, Nigel, ed. *The World in 1984.* Vol. 1. Baltimore: Penguin Books, 1964.

Caldicott, Dr. Helen. *Missile Envy.* New York: William Morrow & Company, 1984.

Cernan, Eugene. *The Last Man on the Moon.* New York: St. Martin's Press, 1999.

Clarke, Arthur C. *The Exploration of Space.* New York: Harper & Brothers, 1951.

———. *The Promise of Space.* New York: Harper & Row, 1968.

———. *Rendezvous with Rama.* New York: Bantam Books, 1990 (paperback).

Comins, Neil F. *What If the Moon Didn't Exist?* New York: HarperCollins, 1993.

De Fontbrune, Jean-Charles. *Nostradamus: Countdown to Apocalypse*. New York: Pan Macmillan, 1986.

Dyson, Freeman J. *Weapons and Hope*. New York: Harper & Row, 1984.

Eckart, Peter, ed. *The Lunar Base Handbook: An Introduction to Lunar Base Design, Development, and Operations*. New York: McGraw-Hill, 1999.

Emme, Eugene M., ed. *Science Fiction and Space Futures*. Vol. 5, AAS History Series. San Diego: Univelt (for the American Astronautical Society), 1982.

Ezell, Linda Neuman. *NASA Historical Data Book*. Vol. 3. Programs and Projects 1969–1978, NASA Historical Series, NASA SP-4012. Washington: National Aeronautics and Space Administration, 1988.

Foundation for the Future. *The Next Thousand Years TV Series Project*. Bellevue, Wash.: 2002.

Francis, Peter. *Volcanoes*. New York: Penguin, 1976.

Gehrels, Tom, Mildred Shapley Matthews, and A. M. Schumann, eds. *Hazards Due to Comets and Asteroids*. Tucson: University of Arizona Press, 1995.

Glasstone, Samuel. *Sourcebook on the Space Sciences*. New York: D. Van Nostrand Co., 1965.

Hanle, Paul A., and Von Del Chamberlain, eds. *Space Science Comes of Age*. Washington, D.C.: National Air and Space Museum, 1981.

Harrison, Albert A. *Spacefaring: The Human Dimension*. Berkeley: University of California Press, 2001.

Kaempffert, Waldemar. *Science Today and Tomorrow*. New York: Viking Press, 1945.

Kieffer, Hugh H., Bruce M. Jakosky, Conway W. Snyder, and Mildred S. Matthews, eds. *Mars*. Tucson: University of Arizona Press, Space Science Series, 1992.

Kranz, Gene. *Failure Is Not an Option*. New York: Simon & Schuster, 2000.

Kurzweil, Ray. *The Age of Spiritual Machines*. New York: Viking, 1999.

Launius, Roger D., and Howard E. McCurdy. *Imagining Space: Achievement, Predictions, Possibilities, 1950–2050*. San Francisco: Chronicle Books, 2001.

LeMay, General Curtis E., with Major General Dale O. Smith. *America Is in Danger*. New York: Funk & Wagnalls, 1968.

Levine, Arnold S. *Managing NASA in the Apollo Era*. NASA History Series. Washington, D.C.: National Aeronautics and Space Administration, 1982.

Lewis, John S., and Ruth A. Lewis. *Space Resources: Breaking the Bonds of Earth*. New York: Columbia University Press, 1987.

Lewis, Richard S. *The Voyages of Apollo: The Exploration of the Moon*. New York: Quadrangle, 1974.

Lindsey, Hal, with C. C. Carlson. *The Late Great Planet Earth*. Grand Rapids: Zondervan, 1970.

Logsdon, John M. *The Decision to Go to the Moon*, Chicago: University of Chicago Press, 1970.

——, ed. *Exploring the Unknown*. Vol. 1. NASA History Series, SP-4218. Washington, D.C.: National Aeronautics and Space Administration, 1995.

Mackenzie, Dana. *The Big Splat: Or How the Moon Came to Be*. New York: John Wiley & Sons, 2003.

Mangold, Tom, and Jeff Goldberg. *Plague Wars: A True Story of Biological Warfare*. London: Macmillan, 1999.

Mark, Hans. *The Space Station*. Durham: Duke University Press, 1987.

Mautner, Michael N. *Seeding the Universe with Life: Securing Our Cosmological Future*. Legacy Books, 2000.

McDonald, Robert A., ed. *Corona: The First NRO Reconnaissance Eye in Space*. Bethesda: American Society for Photogrammetry and Remote Sensing, 1997.

McDougall, Walter A. . . . *the Heavens and the Earth*. New York: Basic Books, 1985.

McGovern, James. *Crossbow and Overcast*. New York: William Morrow & Company, 1964.

Murray, Bruce C. *Navigating the Future*. New York: Harper & Row, 1975.

Newell, Homer E. *Beyond the Atmosphere: Early Years of Space Science*. NASA History Series, SP-4211. Washington, D.C.: National Aeronautics and Space Administration, 1980.

Nolan, Janne E. *The Trappings of Power: Ballistic Missiles in the Third World*. Washington, D.C.: Brookings Institution, 1991.

O'Neill, Gerard K. *The High Frontier: Human Colonies in Space*. Ontario: Apogee Books (Space Studies Institute), 2000.

Petersen, Carolyn Collins, and John C. Brandt. *Hubble Vision*. Cambridge: Cambridge University Press, 1995.

Pringle, Peter, and William Arkin. *SIOP: The Secret U.S. Plan for Nuclear War*. New York: W. W. Norton, 1983.

Ratcliff, John D., ed. *Science Year Book of 1945*. Garden City, N.Y.: Doubleday, Doran & Company, 1945.

Rees, Martin. *Our Final Hour*. New York: Basic Books, 2003.

Rhodes, Richard. *The Making of the Atomic Bomb*. New York: Simon & Schuster, 1986.

Riabchikov, Evgeny. *Russians in Space*. Garden City, N.Y.: Doubleday & Company, 1971.

Ryan, Cornelius, ed. *Across the Space Frontier.* New York: Viking Press, 1952.

Sagan, Carl. *Cosmos.* New York: Random House, 1980.

———. *Pale Blue Dot.* New York: Random House, 1994.

Schell, Jonathan. *The Fate of the Earth.* New York: Alfred A. Knopf, 1982.

Short, Nicholas M., et al. *Mission to Planet Earth: Landsat Views the World.* Washington: National Aeronautics and Space Administration, 1976.

Shternfeld, Ari. *Soviet Space Science.* New York: Basic Books (translation), 1959.

Shute, Nevil. *On the Beach.* New York: William Morrow & Company, 1957.

Sorensen, Theodore C. *Kennedy.* New York: Harper & Row, 1965.

Spector, Leonard S., and Jacqueline R. Smith. *Nuclear Ambitions.* Boulder: Westview Press (Carnegie Endowment for International Peace), 1990.

Taylor, L. B., Jr., and C. L. Taylor. *Chemical and Biological Warfare.* New York: Franklin Watts, 1985.

Truman, Harry S. *Years of Trial and Hope* (*Memoirs*, vol. 2). Garden City, N.Y.: Doubleday & Co., 1956.

Tsiolkovsky, Konstantin E. *Beyond the Planet Earth.* New York: Pergamon Press, 1960.

Tsipis, Kosta. *Arsenal: Understanding Weapons in the Nuclear Age.* New York: Simon & Schuster, 1983.

Tyson, Neil deGrasse. *The Sky Is Not the Limit.* Amherst, N.Y.: Prometheus Books, 2004.

Verger, Fernand, Isabelle Sourbes-Verger, and Raymond Ghirardi. *The Cambridge Encyclopedia of Space.* Cambridge: Cambridge University Press, 2003.

Ward, Peter D., and Donald Brownlee. *Rare Earth: Why Complex Life Is Uncommon in the Universe.* New York: Copernicus, 2000.

Weaver, Warren, ed. *The Scientists Speak.* New York: Boni & Gaer, 1947.

Weissman, Steve, and Herbert Krosney. *The Islamic Bomb.* New York: Times Books, 1982.

Welch, R. W., *Comet of Nostradamus*, St. Paul: Llewellyn Publications, 1951.

Williams, Peter, and David Wallace. *Unit 731.* London: Hodder & Stoughton, 1989.

Wilson, Damon, *The Mammoth Book of Prophecies: The Predictions of Nostradamus and Other Prophets, Visionaries and Seers*, New York: Carroll & Graf, 2003.

Yenne, Bill. *The Encyclopedia of US Spacecraft.* New York: Exeter Books, 1985.

Yeomans, Donald K. *Comets: A Chronological History of Observation, Science, Myth, and Folklore.* New York: John Wiley & Sons, 1991.

## Articles

Ashley, Steven. "Next Generation GPS." *Scientific American*, September 2003.

"Asking for the Moon." *Science*, November 23, 1984.

Baclawski, Joseph A., and Thomas B. Nath. "The Civil Applications Committee's Role in National Reconnaissance." *Bulletin* (Chantilly: Center for the Study of National Reconnaissance [NRO]), 2002.

Bissell, Tom. "A Comet's Tale." *Harper's*, February 2003.

Braun, David. "Is a Large Asteroid Headed for Impact with Earth in 2880?" *National Geographic News* (Internet), April 4, 2002.

Bulban, Erwin J. "Economic Benefits of Lunar Base Cited." *Aviation Week & Space Technology*, April 18, 1983.

Burrows, William E., and Robert Shapiro. "An Alliance to Rescue Civilization." *Ad Astra*, September/October 1999.

Chaikin, Andrew. "Shoot for the Moon." *Air & Space/Smithsonian*, December 1991/January 1992.

Covault, Craig. "Manned U.S. Lunar Station Wins Support." *Aviation Week & Space Technology*, November 19, 1984.

Crouch, Thomas D. "To Fly to the World in the Moon: Cosmic Voyaging in Fact and Fiction from Lucian to *Sputnik*." In Emme, *Science Fiction and Space Futures*.

Deckard, Margo R. "A Technology for a Better Future: Space Solar Power, an Unlimited Energy Source." Chapter in O'Neill, *High Frontier.*

Dupont, Daniel G. "Nuclear Explosions in Orbit." *Scientific American*, June 2004.

Erikson, Ray. "The Spaceways System: Space Infrastructure Development (Spindev) for Promoting Human Spaceflight." *Space Times* (AAS), May 2002.

———. "Why Space Stations Are Crucial to Our Future." *Space Times* (AAS), January/February 2003.

Evans, Robert. "Blast from the Past." *Smithsonian*, July 2002.

Hahn, G., et al. "The EUNEASO Project: A European NEO Search, Follow-up, and Physical Observation Programme." In Remo, *Near-Earth Objects.*

Hale, Edward Everett. "The Brick Moon." *Atlantic Monthly* (Boston: Fields, Osgood, & Co.), 1869.

Handberg, Roger. "Back to the Future: American Human Spaceflight Returns to Its Roots." *Space Times* (AAS), May/June 2003.

Hartmann, William K. "The Paradigm and the Pendulum." *Nature*, April 20, 2000.

Horvath, Joan C. "Blastoffs on a Budget." *Scientific American*, April 2004.

"How Lethal Was the K-T Impact?" *Science*, September 17, 1993.

Joy, Bill. "Why the Future Doesn't Need Us." *Wired*, April 2000.

Keller, Bill. "The Thinkable." *New York Times Magazine*, May 4, 2003.

Kerr, Richard A. "Whiff of Gas Points to Impact Mass Extinction." *Science*, February 23, 2001.

Knight, Jonathan. "US Reflects on Flying Eye for Transgenic Crops." *Nature*, September 11, 2003.

Lawler, Andrew. "Afghanistan's Challenge." *Science*, November 8, 2002.

———. "Buddhas May Stretch Out, If Not Rise Again." *Science*, November 8, 2002.

———. "Mayhem in Mesopotamia." *Science*, August 1, 2003.

———. "Then They Buried Their History." *Science*, November 8, 2002.

Lowman, Paul D., Jr. "Space Exploration and Plate Tectonics." *Space Times* (AAS), July/August 2003.

Matthews, Robert. "A Rocky Watch for Earthbound Asteroids." *Science*, March 6, 1992.

Mautner, Michael N. "Space-Based Genetic Cryoconservation of Endangered Species." *Journal of the British Interplanetary Society* 49 (1996).

Miller, Dennis M., John E. Stocker, and William C. Martel. "Commercial Space Systems: Implications for National Security." *Bulletin* (Chantilly: Center for the Study of National Reconnaissance), 2002.

"The New Race to the Moon." *Science*, May 2, 2003.

"New Technological Approaches—Not Just Better Policies Required." *Update* (New York Academy of Sciences), March 2003.

Nicolaou, K. C., and N. C. Boddy. "Behind Enemy Lines." *Scientific American*, May 2001.

O'Dale, Charles D. "Establishing an Infrastructure for Commercial Space." *Journal of the British Interplanetary Society*, February 1997.

Pauley, Daniel, and Reg Watson. "The Last Fish." *Scientific American*, July 2003.

Penn, Jeffrey. "Explaining the Demise of Dinosaurs (And Other Mass Extinctions)." *Update* (New York Academy of Sciences), March 2003.

Perkins, S. "Killer Crater." *Science News*, March 15, 2003.

Phoenix, Chris, and K. Eric Drexler. "Safe Exponential Manufacture." *Nanotechnology*, June 9, 2004.

Renne, Paul R. "Flood Basalts—Bigger and Badder." *Science*, June 7, 2002.

Reynolds, David West. "The Space Frontier Paradigm: Correcting a Misleading View of Space as the New West." *Space Times*, May/June 2004.

Schweickart, Russell L., et al. "The Asteroid Tugboat." *Scientific American*, November 2003.

Shermer, Michael. "Codified Claptrap." *Scientific American*, June 2003.

"South Korea Admits to Laser Enrichment Program." *Science*, September 10, 2004.

"Space Scientists Support Satellites as Aid for Afghanistan." *Nature*, October 7, 2004.

Stevens, Jane Ellen. "Bumpy Road to Mars." *Smithsonian*, June 2004.

Stone, Richard. "Fears Grow of Nuclear Brain Drain to Iran." *Science*, March 7, 2003.

———. "Nuclear Trafficking: 'A Real and Dangerous Threat.' " *Science*, June 1, 2001.

"Summit Pledges Global Data Sharing." *Science*, April 30, 2004.

Susser, Ezra S., Daniel B. Herman, and Barbara Aaron. "Combating the Terror of Terrorism." *Scientific American*, August 2002.

Teets, Peter B. "The Challenges for National Reconnaissance in the First Decade of the 21st Century." *Bulletin* (Chantilly: Center for the Study of National Reconnaissance), 2002.

Tikhonravov, Mikhail K. "The Creation of the First Artificial Earth Satellite: Some Historical Details." In *History of Rocketry and Astronautics*. AAS History Series, vol. 8, 1989.

Turner, Andrew E. "Aquarius Launch Vehicle: Failure *Is* an Option." *Space Times* (AAS), May/June 2001.

———. "The Ultimate in Data Storage Security: Archiving Data on the Moon." *Space Times* (AAS), July/August 2003.

"Virgin Territory." *New York Times Magazine*, November 7, 2004.

Von Braun, Wernher. "Exploration to the Farthest Planets." *The World in 1984*, vol. 1.

Webb, David C. "The Strange Career of Landsat: Lessons Learned from a Lack of Policy?" *Space Times* (AAS), July/August 2003.

Wheelon, Albert D. "Technology and Intelligence." *Technology In Society*, vol. 26, April–August 2004.

"Whither Arctic Ice? Less of It, for Sure." *Science*, August 30, 2002.

## Reports, Studies, Papers, and Lectures

Allen, R. D. "Early Lunar Base Concepts of the U.S. Air Force." Sunnyvale: Lockheed Missiles and Space Company, 1992. (Presented at the 43rd Congress of the International Astronautical Federation.)

"Analysis of Cost Estimates for the Space Shuttle and Two Alternate Programs," B-173677. Washington, D.C.: General Accounting Office, June 1, 1973.

Becker, Luann, Robert J. Poreda, Andrew G. Hunt, Theodore E. Bunch, and Michael Rampino. "Impact Event and the Permian-Triassic Boundary: Evidence for Extraterrestrial Noble Gases in Fullerenes." *Science*, February 23, 2001.

Brown, P., et al. "The Flux of Small Near-Earth Objects Colliding with the Earth." *Nature*, November 21, 2002.

Chapman, Clark R., Daniel D. Durda, and Robert E. Gold. "The Comet/Asteroid Impact Hazard: A Systems Approach." Boulder: Southwest Research Institute, February 24, 2001.

Cochran, Thomas B., William M. Arkin, and Milton M. Hoenig. *U.S. Nuclear Forces and Capabilities*. Nuclear Weapons Data Book, vol. 1. Cambridge: Ballinger for the Natural Resources Defense Council, 1984.

Committee on Commerce, Science, and Transportation. *National Aeronautics and Space Act of 1958, as Amended, and Related Legislation*. 95th Cong., 2d sess., December 1978.

Correll, Randall R. "National Security Implications of the Asteroid Threat." Washington, D.C.: George C. Marshall Institute, February 4, 2003.

Edelson, Burton I. "Communication Satellites: The Experimental Years." *History of Rocketry and Astronautics*, vol. 12 (AAS History Series). San Diego: American Astronautical Society, 1991.

"Final Report of the Commission on the Future of the United States Aerospace Industry." Arlington, Va.: November 2002.

Forman, Brenda. "Launch Systems: Shuttle, ALS & NLS." Lecture given at the Graduate School of Engineering, University of Southern California, Autumn 1992.

Fudge, Michael L. "LEO Constellation Orbital Debris Threat Assessment: A Case Study." Proceedings of the 18th American Institute of Aeronautics and Astronautics International Communications Satellite Systems Conference, Oakland, Calif., April 2000.

Giorgini, J. D., et al. "Asteroid 1950 DA's Encounter with Earth in 2880: Physical Limits on Collision Probability Prediction." *Science*, April 5, 2002.

Glasstone, Samuel, and Philip J. Dolan. *The Effects of Nuclear Weapons.* 3rd ed. Washington, D.C.: U.S. Department of Defense and the Energy Research and Development Administration, 1977.

Helin, Eleanor F., Steven H. Pravdo, David L. Rabinowitz, and Kenneth J. Lawrence. "Near-Earth Asteroid Tracking (NEAT) Program." In Remo, *Near-Earth Objects.*

Heppenheimer, T. A. *The Space Shuttle Decision.* NASA History Series, SP-4221. Washington, D.C.: National Aeronautics and Space Administration, 1999.

Hoffert, Martin I., et al. "Advanced Technology Paths to Global Climate Stability: Energy for a Greenhouse Planet." *Science*, November 1, 2002.

Keeney, L. Douglas, ed. *The Doomsday Scenario (Emergency Plans Book).* St. Paul: MBI Publishing Co., 2002.

Lowman, Paul D. Jr. "Return to the Moon: A New Strategic Evaluation." Presentation made at the Space Resources Roundtable, Colorado School of Mines, 1999.

Mazanek, D., et al. "Comet/Asteroid Protection System (CAPS): A Space-Based System Concept for Revolutionizing Earth Protection and Utilization of Near-Earth Objects." IAC-02-IAA.13.4/Q.5.1.01. Paris: International Astronautical Federation, October 2002.

"Military Lunar Base Program." Vol. 1, Project No. 7987, Directorate of Space Planning and Analysis. Los Angeles: Air Force Ballistic Missile Division, April 1, 1960.

National Commission on Space. *Pioneering the Space Frontier.* New York: Bantam, 1986.

*The 9/11 Commission Report,* authorized ed. New York: W. W. Norton & Company, 2004.

*The Post-Apollo Space Program: Directions for the Future* (Agnew Report). Washington, D.C.: Space Task Group Report to the President, September 1969.

Remo, John L., ed. *Near-Earth Objects: The United Nations International Conference.* In *Annals of the New York Academy of Science* 822 (1997).

*Report of the Presidential Commission on the Space Shuttle Challenger Accident,* vol. 2. Washington, D.C.: 1986.

Ride, Sally K. "Leadership and America's Future in Space." Washington, D.C.: National Aeronautics and Space Administration, 1987.

Ruffner, Kevin C. *Corona: America's First Satellite Program*. Washington, D.C.: Center for the Study of Intelligence (CIA), 1995.

Segura, Teresa L., Owen B. Toon, Anthony Colaprete, and Kevin Zahnle. "Environmental Effects of Large Impacts on Mars." *Science*, December 6, 2002.

Shevshenko, Vladislav V. "Lunar Base Siting." In Eckart, *The Lunar Base Handbook*.

*The Spaceguard Survey, Report of the NASA International Near-Earth-Object Detection Workshop*. Pasadena: Jet Propulsion Laboratory/California Institute of Technology (for NASA's Office of Space Science and Applications), January 25, 1992.

Spitale, Joseph N. "Asteroid Hazard Mitigation Using the Yarkovsky Effect." *Science*, April 5, 2002.

*Study to Determine the Feasibility of Extending the Search for Near-Earth Objects to Smaller Limiting Diameters, Report of the Near-Earth Object Science Definition Team*. NASA Office of Space Science, August 22, 2003.

Synthesis Group. "America at the threshold." Arlington (1225 Jefferson Davis Highway, Suite 1501): 1991.

Teich, Albert H., Stephen D. Nelson, and Stephen J. Lita, eds. *Science and Technology in a Vulnerable World*. AAAS Publication Number 02-5A. Washington, D.C.: American Association for the Advancement of Science, 2002.

Tikhonravov, Mikhail K. "The Creation of the First Artificial Earth Satellite: Some Historical Details." *History of Rocketry and Astronautics* 8. San Diego: American Astronautical Society, 1989.

Ward, Steven N., and Erik Asphaug. "Asteroid Impact Tsunami of 2880 March 16." *Geophysical Journal International*, June 2003.

Winter, Frank H. "Camera Rockets and Space Photography Concepts Before World War II." *History of Rocketry and Astronautics*, AAS History Series, vol. 8. San Diego: American Astronautical Society, 1989.

## Minutes

"Discussion at the 257th Meeting of the National Security Council, Thursday, August 4, 1955." Eisenhower Library.

# ACKNOWLEDGMENTS

*The Survival Imperative* grew out of a project called the Alliance to Rescue Civilization, which was conceived by Professor Robert Shapiro, a friend and colleague, then in the Department of Chemistry at New York University. Bob's intelligence, insight, and imagination have been inspirational, as has been his unwavering belief in ARC and this book.

Similarly, I was fortunate to have two world-class aerospace engineers as backfield coaches. Ray Erikson could be called a renaissance engineer, given his rare combination of outstanding intellectual capacity, vision, and the soul and literary instinct of a poet. He was there at every stage, making helpful suggestions, steering me back on course when that was required, reading some chapters to catch errors (and in a number of instances, finding them), and extending his friendship. Andy Turner was the other dependable "godfather," who not only provided a model with his own lunar archive, but who encouraged me throughout and, like Ray, read the more technical parts of this story and sent a stream of corrections and insightful observations. (As usual, though, the errors are mine alone.) The "Sources" section is an extension of this acknowledgment, and so is the body of the text, since both Ray and Andy have published their ideas, and I have helped myself to them generously.

Tom Jones, a veteran astronaut, and Steve Wolfe, a veteran of Capitol Hill, have also generously shared ideas for this project and have an enthusiasm about space and the defense of the planet that is infectious. Steve is also one of ARC's founding fathers.

Dr. E. Myles Standish, an old friend and an astronomer at the Jet Propulsion Laboratory, helped set the stage for this story by giving me a primer on the all but ungraspable (for me, at least) distances of the objects that inhabit the cosmos.

Kay McCauley is technically my agent. But that substantially fails to describe the friendship and wisdom that marvelous lady brought to our relationship, much to the benefit of author and book. I have gladly followed her invaluable advice (usually offered at a certain seafood establishment in midtown Manhattan). *Feedback* is an overused expression, but she gave it so intelligently and copiously, it deserves mention here. I am fortunate to have had her, too, there for me.

I am also fortunate that Joelle once again graciously volunteered to read the manuscript for everything from typos to problems with syntax to repetitions (and found prodigious numbers of all of them). She has yet again earned my deep gratitude on that score alone.

Patrick Nielsen Hayden, my editor, also made a number of valuable observations that helped on the literary side. He proved to be exceptionally bright, sensitive, and dedicated to the book's success. In fact, Patrick carries a vision for *The Survival Imperative* that even its author could not have conceived.

And there is also Miss Sophie Soltanian, and her brother, Tommy, to thank for being the ultimate justification for planetary protection and this work, which grew out of it. It is therefore offered to them, as another book was once offered to their great-grandma, with love, respect, and appreciation.

Finally, my eternal gratitude to Lara J. Burrows, M.D., who took the high road. She set an example of altruism, courage, and dedication that has been and will always be my inspiration.

# INDEX

# ABOUT THE AUTHOR

William E. Burrows has written about space since the time of Apollo. He has reported for *The New York Times*, *The Washington Post*, *The Wall Street Journal*, and *The Richmond Times-Dispatch*. He is also a contributing editor at *Air & Space/Smithsonian* and has had articles published in *The New York Times Magazine*, *Foreign Affairs*, *Harper's*, and other publications. His books on space include *Deep Black*, a pioneering investigative work on space reconnaissance; *Exploring Space*, which described the exploration of the solar system; and *This New Ocean*, an award-winning history of the space age that has been called the best work of its kind in English. He is also a professor of journalism at New York University and the founder and director of its graduate Science and Environmental Reporting Program.